City Development: Issues and Best Practices

Editors-in-Chief

Huhua Cao, Geography, Environment and Geomatics, University of Ottawa, Ottawa, ON, Canada

John Zacharias, College of Architecture and Landscape, Peking University, Beijing, China

Claude Ngomsi, United Nations Human Settlements Programme (UN-Habitat), Nairobi, Kenya

The current rate of urbanization is unprecedented and poses enormous challenges for governing bodies. New approaches to development and urban management are needed in the context of globalization and the need for local sustainability. While the developing world itself offers an abundance of lessons, case studies and best practices, these have rarely been positioned as cutting-edge contributions to reformed practices in city development. It is well recognized that the experience of the developed world is an incomplete guide to the new challenges posed by urbanization in the contemporary world.

The "City Development: Issues and Best Practices" book series includes academic research, comparative and applied research, and case studies at the scale of the neighborhood, city, region, nation and supranational levels. This series will offer an opportunity to present the latest academic research and best practices in urban development with the goal of promoting sustainable and inclusive development, learning from the diverse and complementary experiences of rapidly urbanizing areas of the world. Although this book series focuses primarily on the developing world, we intend to include the latest academic research and evolving best practices from developed countries.

The series is intended for geographers, planners, engineers, urban designers, architects, political scientists, sociologists, and economists, as well as policy makers and representatives from government, civil society, industry, etc. who are interested in the developing world. The series will also interest students seeking a foundation in the comparative analysis of key issues facing rapidly urbanizing areas of the world. It will include monographs, edited volumes and textbooks. Book proposals and final manuscripts will be peer-reviewed.

The areas to be covered in the series include, but are not limited to, the following:

- Urban Equity and Inclusivity
- Reforming Informal Settlements
- Climate Change and Adaptation of the Built Environment
- Health Crises and Urban Risk Management
- Privacy, Surveillance, Security and Collective Wellbeing
- Governance and Urban Resilience
- Participatory Policy and the Right to the City
- Indigenous and Ethnic Minority Urbanization
- Mobility and Urban Transformation
- Land Use and Urban Landscapes
- Urban Infrastructure and Transport Systems
- Urban Intelligence and Technologies
- Participatory Budgeting and Community Building
- Industrial Parks and Agro-Processing Zones
- Housing and Land Tenure Issues

Liqin Zhang · Elizabeth Kanini Wamuchiru ·
Claude A. Meutchehe Ngomsi
Editors

The City in an Era of Cascading Risks

New Insights from the Ground

Editors
Liqin Zhang
Department of Land Resources Management
China University of GeoSciences (Wuhan)
Wuhan, Hubei, China

Elizabeth Kanini Wamuchiru
Department of Urban and Regional Planning
University of Nairobi
Nairobi, Kenya

Claude A. Meutchehe Ngomsi
Regional Representation for Africa
UN-Habitat
Nairobi, Kenya

ISSN 2731-7773 ISSN 2731-7781 (electronic)
City Development: Issues and Best Practices
ISBN 978-981-99-2049-5 ISBN 978-981-99-2050-1 (eBook)
https://doi.org/10.1007/978-981-99-2050-1

© The Editor(s) (if applicable) and The Author(s), under exclusive license to Springer Nature Singapore Pte Ltd. 2023

This work is subject to copyright. All rights are solely and exclusively licensed by the Publisher, whether the whole or part of the material is concerned, specifically the rights of translation, reprinting, reuse of illustrations, recitation, broadcasting, reproduction on microfilms or in any other physical way, and transmission or information storage and retrieval, electronic adaptation, computer software, or by similar or dissimilar methodology now known or hereafter developed.
The use of general descriptive names, registered names, trademarks, service marks, etc. in this publication does not imply, even in the absence of a specific statement, that such names are exempt from the relevant protective laws and regulations and therefore free for general use.
The publisher, the authors, and the editors are safe to assume that the advice and information in this book are believed to be true and accurate at the date of publication. Neither the publisher nor the authors or the editors give a warranty, expressed or implied, with respect to the material contained herein or for any errors or omissions that may have been made. The publisher remains neutral with regard to jurisdictional claims in published maps and institutional affiliations.

This Springer imprint is published by the registered company Springer Nature Singapore Pte Ltd.
The registered company address is: 152 Beach Road, #21-01/04 Gateway East, Singapore 189721, Singapore

Preface

The International Conference on Canadian, Chinese and African Sustainable Urbanization (ICCCASU) is pleased to present the first Springer Book Series: "City Development: Issues and Best Practices", and the first book of the series: "Cities in an Era of Cascading Risks: New Insights from the Ground"; a 2021 cooperative venture with Springer Publishing. Rapid urban growth in developing countries and the recent experienced global pandemic have called for a new way of looking at sustainable development with a focus not only on protecting the earth's resources but also ensuring human health.

ICCCASU, a joint initiative, was created between the United Nations Human Settlements Programme (UN-Habitat) and the University of Ottawa in 2015. Since its inception the organization has expanded to include a consortium of several Canadian, Chinese and African universities, NGOs, and it continues creating partnerships in various regions around the world. As an international think-tank, ICCCASU brings together researchers, representatives from government, civil society, industry and academia for in-depth discussions on cities and urbanization to promote sustainable and inclusive urban development in a forum based on the diverse but complementary experiences of Canada, China and African nations. Moving beyond the conventional practice of North-South or South-South cooperation, ICCCASU fosters a triangular dialogue between African countries, Canada and China, which account for more than 30% of the world's urban population.

To date, ICCCASU has mounted four conferences and training programmes. ICCCASU I was held in Ottawa, Canada, in 2015, while in conjunction with the government of Cameroon, ICCCASU II was held in Yaoundé, Cameroon, in 2017 (www.icccasu2017.org). In the summer of 2019, ICCCASU 3 was held in Chengdu, China, in partnership with the China Centre for Urban Development (CCUD), affiliated with the Chinese government's National Development and Reform Commission (www.icccasu2019.org). Funded by SSHRC, the fourth conference in Montreal, Canada, in July 2021 marked a significant milestone as the first in the second rotation of discussions among the three regions of interest. Moreover, ICCCASU 4 was the only UN-sponsored conference in Canada that re-examined cities in the COVID-19 context (www.icccasu2021.org). In addition to the biennial conferences,

ICCCASU also hosted a roundtable on Canada's participation in Habitat III in June 2016. ICCCASU was also one of several organizations selected to mount workshops at the 10th World Urban Forum in February 2020 in Abu Dhabi, which carried out policy dialogue on culture and innovation in city-building based on case studies from different regions and perspectives (WUF-ICCCASU Workshop—ICCCASU 2021). At the 11th World Urban Forum in June 2022 in Katowice, Poland, ICCCASU organized a roundtable on "City transformation among socialist and post-socialist countries in the Pandemic context: What is the future?" (https://icccasu2021.org/news/).

Taking advantage of its rich networks, ICCCASU is expanding to conduct transnational and multi/inter-disciplinary empirical research projects and best practices cases on urban development. The team continues to develop various proposals while collaborating with academics, professionals, and politicians worldwide. Our up-to-date research provides best practices for sustainable and inclusive urban development globally, particularly in the developing world.

Many scientific and policy papers presented at ICCCASU conferences and workshops are published in academic journals. These publications have become essential references for professionals and policymakers working in sustainable urban development across the globe. Furthermore, ICCCASU produced a series of reports for each conference and workshop (ICCCASU Reports, 2015, 2017, 2018, 2020), including a study for Global Affairs Canada entitled Recommendations on Areas of Possible Collaboration between Canada and China in Africa (https://icccasu2021.org/reports/). With the wealth of knowledge and network, ICCCASU ventured in 2020, to sign an agreement with Springer to create a new Springer Book Series: "Cities in an Era of Cascading Risks: New Insights from the Ground" (https://icccasu2021.org/springer-book-series/).

ICCCASU's Springer Book Series present the latest academic research and best practices in urban development to promote sustainable and inclusive growth, learning from diverse and complementary experiences of rapidly urbanizing world areas. Sustainable urbanization is relevant today as the rate of urbanization is unprecedented and continues to pose enormous challenges for governing bodies. New approaches to development and urban management are needed in the context of globalization parallel to local sustainability. While the developing world itself offers an abundance of lessons, case studies and best practices, these have rarely been positioned as cutting-edge contributions to reformed practices in city development. It is well recognized that the experience of the developed world is an incomplete guide to the new challenges posed by urbanization in the contemporary world.

The book series includes academic research, comparative and applied research, and case studies at the scale of the neighbourhood, city, region, nation and supranational levels. Although this book series focuses primarily on the developing world, we also aim to include the latest academic research and evolving best practices from developed countries. Geographers, planners, engineers, urban designers, architects, political scientists, sociologists, economists, policymakers and representatives from government, civil society, industry, etc. interested in the urban issues in the developing world will find this book series not only appealing, but also an enriching cutting edge and global research experience. The series will also interest students seeking a foundation in comparative analysis of critical issues facing rapidly urbanizing world areas. The peer-reviewed book proposal and final manuscripts will include monographs, edited volumes and essays.

We are very excited that this first book of our Springer series entitled "The City in an Era of Cascading Risks: New Insights from the Ground" is published after two arduous years of global pandemic. As the world struggled to respond to the SARS-2 COVID-19 virus, cities and its citizens have started to reflect on best urban practices. Some cities proved to be more resilient than others, and this greatly depended on the city's socioeconomic and political position. To validate this hypothesis, this book provides unique perspectives into newly changed political and socio-economic urban landscapes due to COVID-19 in diverse cities, using a global perspective in a post-pandemic era.

This book is divided into three sections with fifteen chapters overall. It explores the impacts of the COVID-19 on city planning, building and maintenance; it considers city resilience and what risks cities are facing, and it examines urban development from diverse socioeconomic and political perspectives. The book contains multidisciplinary work by authors from China, Angola, Burundi, Cameroon, Kenya, Morocco, Nigeria, Canada, Italy, Poland and France. Many of the papers consist of either a case study, comparative study, or a research study of individual cities across the globe, some compared and contrasted. Both citizen and city planner perspectives are considered throughout the book, seeking out as many ways as possible to improve urban city resilience while expanding the book's readership. Qualitative and quantitative data collection methods are used, with many studies conducting surveys and over-the-phone or/and online interviews to gather sufficient data. At the same time, secondary research was also being undertaken, such as literature reviews, in some cases before a study was conducted, to collect existing information on city resilience, urban development and planning, city management, globalization, healthy cities, and informal settlements, and many other topics. Mobility is an essential aspect of the book, discussing how it plays a prominent role in the social segregation of a city, the city's planning, the urban city's resilience and the changing mobility habits due to travel restrictions because of COVID-19.

A post-pandemic understanding of the new and constantly shifting socioeconomic and political landscapes is in high demand as the world transitions out of the emergency restrictions from the COVID-19 pandemic. Understanding how cities responded and reacted to COVID-19 can offer insight into urban management and resource allocation in future and current pandemics, allowing metropolitan towns to become more resilient. Lessons concerning the resilience of cities can be learned and applied to and from individual cities' urban city development and governance, encouraging information exchanges that benefit all parties involved.

This book offers a unique post-pandemic perspective of urban cities as it is in both current and forecasted demand and literature discussing post-pandemic views are limited, provides a global perspective to improve the resilience of metropolitan cities, especially in the post-pandemic era and brings together a diverse group of scholars from across the global south and north to examine cities across the world to improve urban cities.

We are also delighted to let our readers know that the second book, "Making Sense of Planning and Development for the post-pandemic Cities", is prepared and will be available soon (see the details on Call for contributions at https://mailchi.mp/b60a018a8998/icccasu-newsletter-april2022?e=b9e79d6154).

Over the past several years, ICCCASU has expanded its activities from conferences to training/workshops, consultations, publications, research projects and best practices cases. Additionally, it has proved to be a reliable and authoritative voice, developing networks among urban researchers and practitioners worldwide. Cities are centres for innovation. When managed sustainably and inclusively, urban development can lift people out of poverty, improve access to education, foster economic growth and reform land management. With a track record of significant contributions to the crucial discussions on sustainable and inclusive urban development, ICCCASU's reputation as an international think-tank is well established. To that end, ICCCASU 5 will be held in 2023 in one of the African countries.

Wuhan, China	ICCCASU Organization Committee
Nairobi, Kenya	Liqin Zhang
Nairobi, Kenya	Elizabeth Kanini Wamuchiru
	Claude A. Meutchehe Ngomsi

Contents

Part I Impact of COVID-19 Pandemic in Cities

1 Appraising the Impacts of COVID-19 and Climate Change on Urban Residents 3
Abimbola Omolabi

2 Beyond COVID-19: Planning the Mobility and Cities Following "15-Minute City" Paradigm 25
Tiziana Campisi and Kh Md Nahiduzzaman

3 Youth-Led Response to COVID-19 in Informal Settlements: Case Study of Mathare, Nairobi, Kenya 37
Douglas Ragan, Olga Tsaplina, Amandine Mure-Ravaud, Mary Hiuhu, Isaac Muasa, and Kui Cai

4 COVID-19 in Angola's Slums: Providing Evidence and a Roadmap for Participatory Slum Upgrading in Pandemic Times 53
Miriam Ngombe, João Domingos, and Allan Cain

5 Corporate Social Responsibility (CSR) Response to COVID-19 in Africa: Towards Healthy Cities 75
Raynous Abbew Cudjoe

Part II Resilience and Risk in Contemporary Cities

6 Southern Cities Between Rapid Urbanization and Increasing Need for Flood Mitigation Measures, The Case of Bujumbura, Burundi 95
Gamaliel Kubwarugira, Mohammed Mayoussi, and Yahia El Khalki

7 The Heritagisation of Social Housing, a Model of Ordinary Urban Resilience? 107
Géraldine Djament

8 **The City-State of Hamburg (Federal Republic of Germany): Metropolization, Port Functions and Urban Resilience** 129
Patricia Zander

9 **Small Towns Ageing—Searching for Linkages Between Population Processes** 149
Jerzy Bański, Wioletta Kamińska, and Mirosław Mularczyk

Part III Urban Development from Diverse Perspectives

10 **China's Integrated Urban–Rural Development: A Development Mode Outside the Planetary Urbanization Paradigm?** 169
Chen Yang and Zhu Qian

11 **The Transport System as a Process of Territorial Development or the Risk of Social Segregation** 195
Riad Arrach, Amal El Jirari, and Oussama Benmakrane

12 **Location Choices of Micro Creative Enterprises in China: Evidence from Two Creative Clusters in Shanghai** 213
Zhu Qian

13 **On the Spatial Formation Mechanism and Inclusive Development of Tibetan Commodity Streets in Chengdu City** 231
Xiao Qiong

14 **Rescaling of Chinese Urban Space: From the Perspective of Spatial Politics** 243
Xiaoxi Liu and Qianning Li

15 **Global Cities and Business Internationalization: Towards a New Interdisciplinary Research Agenda** 259
Abdelhamid Benhmade, Philippe Régnier, and Martine Spence

About the Editors

Liqin Zhang is a researcher at the University of Ottawa. She was a professor at China's University of GeoSciences (China), expertizing land cover change, land use planning and management, as well as tourism plan and management. She received her second Ph.D. in urban geography at the University of Ottawa (Canada), specializing in urban and regional studies, particularly on topics related to urbanization, social ecology, urban land growth and sustainability. Dr. Zhang has led and co-led over forty transdisciplinary research and training projects funded by Canada and China. She published many articles and chapters in well-known academic journals and professional books. She works as co-founder and publication chair of ICCCASU (www.icccasu20 21.org). She is currently the editor-in-chief and founder of Sci-Hall Press, an open access academic publisher (www.sci-hall.com).

Elizabeth Kanini Wamuchiru Ph.D. is currently a lecturer at the Department of Urban and Regional Planning, University of Nairobi, Kenya. She is a trained and experienced urban and regional planner both in academics and in practice. She holds a degree in Urban and Regional Planning (University of Nairobi); Master degree in Urban and Regional Planning (University of Nairobi); Master degree in Human Settlements (KU Leuven, Belgium); and a Ph.D. in Planning (TU Darmstadt, Germany). Her research interests mainly focus on urban infrastructure systems, mobility studies, community participation, urban housing, informality and inequality debates. Her recent research focuses on policy-practice agenda for disability-inclusive urban transport system in Accra and Nairobi and impacts of the coronavirus pandemic in the Global South.

Dr. Claude A. Meutchehe Ngomsi is the Regional Adviser in charge of the Central Africa Region, Mauritania, Guinea and Madagascar as Programme Management Officer at the Regional Office for Africa (since January 2020). He previously served as Human Settlements Officer in charge of Francophone countries in Africa (November 2015–December 2019). He has more than 20 years of experience in international urban development, housing, and crime and conflict prevention. He led the formulation of the Rwanda National Urbanization Policy between 2013 and 2015, and the modernization of the Ouagadougou Municipal Police by establishing the Urban Safety Observatory in Ouagadougou, in Burkina Faso (2010 to 2013). He holds a Ph.D. in Urban Geography and Crime Prevention (University of Yaoundé I) and a master's in Human Geography.

Part I
Impact of COVID-19 Pandemic in Cities

Chapter 1
Appraising the Impacts of COVID-19 and Climate Change on Urban Residents

Abimbola Omolabi

1.1 Introduction

Urbanization described as the gradual shift in human population residence from rural to urban area occurs in a manner that increases the proportion of a population living in cities than rural areas, growth in urban population size and area which bodes well for the location of administrative facilities and functions [1]. The phenomenon results in a situation where 3% of the earth surface occupies urban areas, and 55% of the world population lives in urban areas with an estimated increase to 70% by 2050 [2]. This clearly reinstates to urban experts that the urban age has finally caught up with mankind. Nigeria's situation is not an exception.

The world Bank Report of January 2022, reported that the Nigerian Urban Population in 2020 was 107,106,007, constituting (51.96%) of the total population, and by the year 2050, the country's urban population figure will hit 296,480,000 representing 67.1% of the total population. Thereby categorizing Nigeria, India and China among the examples of few countries that would account for 35% of urban world population between 2018 and 2050 [3, 4]. The trend of Nigerian urbanization depicts an interesting pattern which should be managed effectively towards building a sustainable resilient city.

Factors which have contributed to Nigeria's accelerated rate of urbanization and urban growth include natural increase, and rural–urban migration among others. As the country's urban areas swell, strong city planning is vital in managing the problems and difficulties such as overcrowding which occurs when the number of individuals exceeds the available space and lacks access to adequate infrastructural facilities. In the contemporary world, it has been observed that some citizens of the developed and developing countries have been living with the consequences of a double integrated hazard of flooding and COVID-19 pandemic that derived

---

A. Omolabi (✉)
Department of Urban and Regional Planning, Yaba College of Technology, Yaba, Lagos, Nigeria
e-mail: bimboomolabi@yahoo.com

© The Author(s), under exclusive license to Springer Nature Singapore Pte Ltd. 2023
L. Zhang et al. (eds.), *The City in an Era of Cascading Risks*, City Development: Issues and Best Practices, https://doi.org/10.1007/978-981-99-2050-1_1

from climate change, urbanization and environmental degradation [5, 6]. Indeed, studies have revealed the co-location of flood hazard and COVID-19 infection that manifested in cities of both the developed and developing countries. For instance, the experiences of developed countries including the United States, China, Japan, and the United Kingdom were reported by Zinda et al. [7], Simonovic et al. [6], Izumi et al. [8]. While the experiences of developing countries such as El Salvador, Lesotho, Kenya; Cameroon, Nigeria, Peru were documented by McKernan and Weichelt [5], Phillips et al. [9], Walton [10].

Attempts to manage these multiple hazards and their nature have taken different approaches that leave much to be desired. Some of the measures taken to combat the menace sometimes are conflicting practically. Simonovic et al. [6] emphasized that community's use of sand bag strategy to flood control was more problematic to achieve during social distancing and social isolation. Similarly, the lock down imposed for the COVID-19 pandemic adversely slowed down people's attempt to relocate to a safer environment from flooding hazards. There was a need to balance community residential environmental quality with urban growth in the face of rapid urbanization, flooding hazard that was dynamic and tended to change with climate change and the COVID-19 pandemic which had been globally responsible for the losses of human life that occurred recently. Ramalho and Hobbs [11] had noted earlier that in order to trade-off between urbanization, community residential neighbourhood quality and human health, it was essential for developing countries to strengthen disaster resilience and achieve sustainable development. The coincidence of urban flooding and COVID-19 hazards as demonstrated recently tended to exacerbate disasters that required a paradigm change in its management. Previous experience suggested that pre-existing flood disaster management had often dealt with the disaster management in an isolated manner without any recourse to pandemic happening simultaneously. The recent experience of overlapping disasters of flooding and COVID-19 reinforced the fact that a new approach was needed in dealing with multiple hazards. The new approach in disaster management that was action based required a move from disaster vulnerability to disaster resilience. It hinged on the understanding that disasters did not impact everyone the same way and that the challenge associated with sustainable human well-being called for a paradigm shift in dealing with multiple hazards.

This background provides the basis for understanding of the practical links that exist between the factor of multiple hazard conditions in a high density residential community and the creation and amplification of disasters that require a paradigm change. Implying the need for considering the adoption of integrated disaster management and sustainable development that would result in mitigation of disaster and reinforcing resilience as a new paradigm.

1.2 Statement of Problem

Cities because of their nature as centers of economic growth, civilization and innovation are homes to most populations of the world with the hope for a better living, social and economic conditions. In a situation where residential communities in cities become overcrowded as a result of growth without ensuring access for adequate infrastructural facilities to health care, water, sanitation and hygiene, often exposed the residents to high risk of transmission of diseases and make them vulnerable to hazards that tend to compromise their structure and in turn their resilience [12, 13]. Vulnerability is the degree to which a system is likely to experience harm due to hazards or may react during the occurrence of a hazardous events during occurrence [14], UNIDSR 2012 cited in [15]. Urban flooding, and COVID-19 pandemic are hazards that have taken its toll resulting in adverse health outcomes in high density urban areas [16–18].

In the immediacy of the pandemic, the group who suffer most and are worse hit are the low-income group and the urban poor [17, 19, 20]. Mushin, the study area is a low income and high density residential community in Lagos that is affected by flooding and has the third highest number of infection cases of COVID-19 as at 10th May 2020 [21–23]. Heretofore, the pre-existing disaster control plans in the study area often deal with the individual disaster scenarios of flooding with a view to minimizing immediate loss of life, not taking account of a pandemic occurring simultaneously. The control plans over the years did not consider the coincidence of the hazard simultaneously, it is proactive in nature and paid lip service to the possibility of future occurrence. There is a need to create a resilient community using the coincidence of the hazards within the urban space and the conceptual framework of sustainable development.

This study therefore appraises the adverse impacts that arise from the complex confluence of the hazards of flooding and COVID-19 disease and operationalizes the concept of resilience as a solution to effectively addressing the multi-hazard problems in an urban space with a view to enhancing the well-being of residents and the potential ability to withstand the immediate and longer-term effect of the disasters.

1.3 Aim and Objectives

The study aims to appraise the impacts of the confluence of urban flooding and COVID-19 multi-hazard problems on the well-being of residents of a crowded community. The objectives are to balance urban growth with the environment while preventing disaster from multi-hazard problems towards the creation of resilient communities.

1.4 Literature Review and Conceptual Issues

1.4.1 Literature Review

Climate Change and Urban Flooding Experience in Lagos

Mankind from creation has depended on the environment for existence and sustenance with adverse impact on the quality of the environment. It has been observed globally that climate change affects the social and environmental determinants of health. Various studies in recent time have established that irrespective of location, climate change is one of the most serious environmental stressors that threatens human well-being [24–26]. The most acceptable definition of climate change by Intergovernmental Panel on Climate Change according to Ilevbare [26], states that 'climate change is the change in the state of the climate that can be identified by using statistical test by changes in the mean and variability of its properties and that persists for an extended period typically decades or longer'.

The phenomenon occurs when a change is observed in Earth's climate which results in a new weather pattern that subsists in a place for an extended period of time [27]. Evidence of climate change situated in literature revealed that the average global temperature has risen to around 1.0 °C or 1.8 F since 1900 with over half of the increase occurring since the mid-1970s. This event results in increased heavy rainfall, sea level rise and snowfall events which increased the risk of flooding [28]. Several studies on Lagos experience confirmed a projection of sea level rise of between 0.32 and 1.41 m in conjunction with increased flooding risk probability by the year 2100 [29, 30] (Odjugo 2001, cited in [31], p. 6).

Views expressed by different authors on the factors that caused climate change resulting in increased flooding are natural and anthropogenic-induced man-made activities that included urbanization, industrialization, deforestation and poor waste disposal management [32, 33] (LASG 2012). Urban flooding has become a major natural hazard in Lagos since 1947 due to growing population, rapid urbanization and extreme weather events [16, 32, 34–38]. The increased flood events caused by climate change coupled with lack of coping capacity, high levels of vulnerability of people, that is aggravated by poor infrastructural and environmental conditions resulting in adverse impact on the life of people.

This provides the basis of argument that access to adequate urban infrastructure constitutes the road map for success in combating climate change induced urban flooding and COVID-19 impact on the well-being of people. In other words, the convergence of multi hazard problems and urban infrastructural inadequacies resulted in devastating ways that could only be reversed by a different approach that is linked with resilience.

Urban resilience is being emphasized as the approach that recognizes stakeholder's fundamental right to participate in the process of capacity building of their community following their exposure to the confluence of multi hazard problems of flooding caused by climate change and COVID-19 pandemic [12, 39].

COVID-19 Pandemic Related Experiences in Lagos

The World Health Organization (WHO) declared Coronavirus Disease 2019 (COVID-19) as a pandemic on March 11th 2020 consequent upon its speed, nature and scale of transmission of the disease [40]. The pandemic spread like fire from countries to countries globally. It was recognized as a highly transmittable and pathogenic viral infection caused by sense acute respiratory syndrome (Bovey 2020). Cities in general have been recognized to be the frontlines of COVID-19 with the urban dwellers facing the devastating health and economic impact of the pandemic due to a complex combination of factors that include population size, density and pressure on public health facilities during the response time [17].

The high concentration of people and activities in Lagos city without adequate infrastructure compounded the vulnerability tendency of the highly densely populated and crowded community to the pandemic. Paradoxically, these same factors are significant for the actions that are needed to mitigate the impacts, improve the well-being of people and enhance urban pandemic resilience. Indeed, within the first six months of the global experience of the pandemic in Nigeria, Lagos recorded the highest figure of COVID-19 pandemic with the index case of 214 out of 373 cases among 20 other states [23]. A situation that required urgent attention towards mitigating its impact on the residents 'well-being'. The government's efforts towards the control of pandemic spread are in tandem with the WHO directive in terms of social distancing, isolation, washing of hands and use of sanitizers (WASH) and mobility restrictions among others.

However, the high concentration of people and activities in the community made the residents' increased vulnerability to the pandemic more worrisome. It is worthy of note that various responses to the spread of COVID-19 disease presupposed that a more robust pragmatic action was needed to minimize the impacts and enhance quality of life of residents of the community. The notion concurred with Milhann (2020) and Connolly et al. [41] postulations that neighbourhood residential quality within the city's spatial territory of the developing countries would be critical for their recovery and restoration from the COVID-19 pandemic and flooding. Thus, reinforcing the fact that post-COVID-19 disaster urban planning in Nigeria needs to leverage on the urban resilience approach for the recovery, and revitalization of high residential density neighbourhood that are adversely affected by the convergence of the climate crisis and COVID-19 devastation.

1.4.2 Conceptual Framework

Neighbourhood Residential Quality, Urban Resilience and Well-Being Nexus

A key factor that plays a significant role for the well-being of the inhabitants of a community is the residential neighbourhood quality. The notion reinforces the paradigm shift of housing research from the concentration on physical attributes of single dwelling units to the consideration of neighbourhood quality as an integral part of dwelling units study in the contemporary resilient cities research. Neighbourhood as an element of the housing environment suggests that access to a good housing environment represents an everyday landscape which can either enhance or limit the physical and social well-being of the residents in a community. It influences health, sanitary condition and productivity of individual; reflects the dimensions of infrastructure resilience that relate to resident's well-being and influenced by a range of factors including access to facilities and services [42–44].

A normative description of residential neighbourhood quality entails level of acceptability of dwelling unit in relation to the functionality of the housing unit, adequacy of housing services, basic amenities, space and sanitary condition, that make the units able to respond to a range of human needs, satisfaction and well-being [45, 46] (Godwin et al. 2020). Being an important component of a resident's meaningful life, the assessment of neighbourhood quality in a high residential density community has a significant important implication for policy making that aims to improve living conditions and well-being of residents in urban areas within the framework of urban resilience.

It is worthy of note that well-being which mostly depends on community level factors reinforces the need for housing policies that provide support for vulnerable group in order to improve their well-being and manage the coincidence of flooding and COVID-19 pandemic hazards [47–49].

Urban Resilience

Urban resilience is described generally as the capacity of the urban system to respond to three converging mega trends that challenge its impact on the well-being of people in the world today. The trends include climate change, urbanization and globalization. It is defined as the measurable ability of an urban system, or community, or society with its inhabitants exposed to hazard to survive, resist, absorb, accommodate, recover, grow and maintain continuity from the effects of a hazard in a timely and efficient manner regardless of what chronic stresses and acute shocks, impact, frequency or magnitude they experience while positively adapting and transforming towards sustainability [50].

From the scholarship and practice perspective, Meerow et al. [51] defined it as the ability of an urban system and all its constituent socio-ecological and socio-technical networks across temporal and spatial scales to maintain or rapidly return to desired functions in the face of a disturbance, to adapt, to change, and to quickly transform systems that limit current or future adaptive capacity.

Contextually, urban resilience is conceived as a fundamental approach that has the potential to respond to the confluence of urban flooding and COVID-19 pandemic problems in the high-density area. Its appropriateness as a strategy for managing the potential hazard in the community derives from the view point of its adequacy in preparation for defense, involvement and responsibility of actors, stakeholders and general populations with actions, plans and community structure that create awareness of the risk and the reasons for action that must be taken. This view corresponds with Prasad et al. [52] postulation of the added human capacity advantage that it has as a social system to anticipate and plan for the future.

The main characteristics of resilience that elaborates its adaptive nature to disaster management is connected with its robustness in terms of flood protection measures and reference to ability to provide uninterrupted services in the event of disruption. This is in addition to its potential capacity to utilize materials to establish, prioritize and achieve goals. The rapidity of its capacity to return the urban system to pre-hazard level of functioning as quickly as possible also reinforces its adaptive capacity to dynamic performance of the urban system under disturbance [53].

Urban resilience futuristic pragmatic application to disaster management as postulated by Willner et al. [54] reifies its relevance and adaptation for studies and analyses of the community under investigation. The four main domains of urban resilience which the study focused on cut across urban landscaping; infrastructural adequacy and functionality; residents empowerment with respect to the capacities to recover, thrive and innovate; and institutional capacity and governance restructuring in such a manner that enhanced stakeholder's collaboration in decision-making and behavioural change towards disaster risk reduction and management at the community level.

Well-Being

Resident's well-being evaluation is defined within the framework of the quality of a community that is viewed as a set of environments for people to interact and use with great influence on their health [55, 56]. The underlying philosophy behind the view presupposes that the quality of a community environmental system and the presence of environmental hazards are adaptable for the description of the residential neighbourhood environment that determines the well-being of residents [57]. It also relates to place making and space creation which hinged on the premise that well-being is enhanced by individual access to a residential neighbourhood quality environment that is influenced by availability and functionality of services and amenities within the community [58].

The implication is that a high density residential area lacking in infrastructural facilities as a geographic space with a defined territory tends to compromise the well-being of residents. The concept synchronized with Garret [59] and Glasser [60] studies that established a significant correlation between high urban density and well-being exacerbated by overcrowding. Contrastingly, Chowell et al. [61] and Hamidi et al. [62] studies observed that high urban density may not necessarily affect the well-being of residents adversely where the high quality residential neighbourhood subsist. Bond et al. [63], Opoko et al. [64] and Ma [65] assertion that studies on the relationship between well-being and residential neighbourhood environment was more common in the developed countries strengthened the need to undertake such a study in the developing countries.

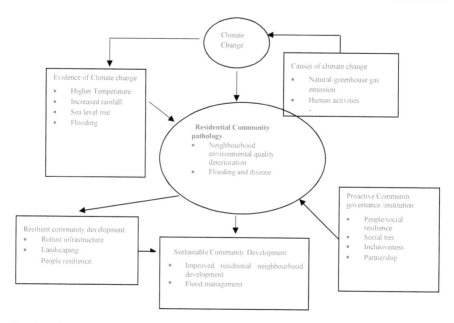

Fig. 1.1 Conceptual framework. *Source* Author (2021)

The experience of coincidence of urban flooding and COVID-19 pandemic hazard in the developing countries had reinforced the necessity and importance of improved residential neighbourhood environmental quality and related rights to infrastructural services as crucial element of cities resilience both in response to pandemic and climate induced disaster [5, 39, 63, 66, 67]. See Fig. 1.1.

In its simplest form, urban revitalization and retrofitting approaches targeted at disaster management bode well for cities' resilience. Its adaptation at addressing urban pathology index at the neighbourhoods level sharpened the competitive edge of an urban area for improved housing quality and resident's well-being [68]. Vidgor's [69] assertion that a cautious revitalisation and retrofitting strategy implementation tend to enhance urban resilience in a highly densely populated and overcrowded area foretell improved well-being of residents.

1.5 The Study Area

The study area is Mushin in Lagos one of the 774 Local Government Areas in Nigeria and located some 535 km South Western part of Abuja the country's capital. Its geographical coordinates are Latitude 60 31′45″ North of the Equator and Longitude 30 21′18″ East of Greenwich Meridian. It is 10 km North of the Lagos city core, and adjacent to the main road to Ikeja the state's capital. Mushin covers an area of 17.01 km², and an estimated population of 1,312,517 and inhabitants are mostly

Yoruba people. The area is densely populated, with mixed uses of commercial and largely congested residential area of 17.01 km² and population density of 51, 161 in 2016 which is far above the Lagos City Density. It is an area that is situated in the flood communities of Lagos and its high residential density exacerbated by high population which poses a threat to the accomplishment of urban planning goals. The inadequate sanitation and drainage facilities, low housing quality and the background of the site which manifests slum development with serious health implications underscores its vulnerability to hazards particularly during a health crisis era aggravated by urban flooding hazard [34, 70, 71].

1.6 Methodology

The research adopted a case study approach and asserted that by virtue of its locational features, residents were at a higher risk of vulnerability than others and had their wellbeing compromised by the coincidence of urban flooding caused by climate change and COVID-19 pandemic problem. More so, vulnerability of a community to natural hazard has often been found to be affected by both their physical and social exposure in addition to preparedness prior to an event, resilience and response following the event [72].

The study site in terms of geographical scope is a local government area that was carefully selected for the purpose of the study because of its background which underscores its vulnerability to hazards. Mushin LGA has the third highest number of COVID-19 infection cases among the twenty LGA in Lagos as at 10th May, 2020 [73]. The study in line with Harwell [74] notion on the research design procedure analyzed collected data using qualitative and quantitative techniques, in addition to the interpretation and presentation of data that are in line with the objective of the case study research.

The sample size generated was 399 using Yamane's (1967) formula cited in Israel [75]. The study site delineation to four zones was based on the administrative parameter, and geographical features aided residential area sampled site selection. A judgmental sampling technique was then adopted to choose a sample from each of the zones. Babalosa, Idi-oro/odi-olowo, Idiaraba and Ojuwoye were selected as the four residential areas for the administration of the questionnaire. These areas depicted clearly the features of Mushin earlier described with housing characteristics and environmental features. A total of 25 questionnaires were randomly administered to the first available and willing 25 households head in four randomly selected streets in each of the four neighbourhoods, and analysis of data was based on a total of 300 (75%) fully completed and returned questionnaires.

The household survey provided information on the aspects of the research geared towards achieving the study's objective. The use of mean score analysis on housing quality assessment, impact of flooding COVID 19-on residents, and appraisal of effectiveness of Government COVID-19 control strategy reflected the quantitative aspect of the study. It was done based on Likert scale rate of 1–5 where 1 is very low and 5 very high in tandem with Canny [76] conception. The descriptive analysis of data collected on socio-economic of respondents, flooding occurrences, causes and impacts, COVID-19 pandemic impact, coping strategies and effectiveness of government strategy on the control of flooding hazard was presented using frequency and percentages.

1.7 Data Presentation and Findings

1.7.1 Socio-Economic Characteristics

Information on socio-economic characteristics is relevant to the study because it relates with residential environmental quality and correlates with the coping capacity of people in the event of a potential adverse effect of impacts of hazards. The analysis of the educational attainment of the respondents revealed that 92 (30%) acquired primary education, while 153 (51%) attained secondary education and 36 (12%) of the respondents had tertiary education. This indicated that the respondents were fairly educated and could appreciate the impacts of the hazard on their life. Their literacy level could be germane to the issue of awareness, regarding the deadly nature of pandemic, preparedness; prevention and mitigation strategies against its spread and future occurrence.

Categories and distribution of occupation among the residents of a community as a major socio-economic attribute often correlates with the income level of an individual while evaluating the resident's affordability of housing facilities, maintenance and coping capacity. Analysis of the respondents' occupations indicated that 135 (45%) of the respondents are traders, 15 (5%) are civil servants, while 60 (20%) and 80 (26.61%) are factory workers and artisans respectively. Only 10 (3.1%) are unemployed. Deductively, the highest percentage of the residents are engaged in informal sector activities in order to eke out a living. This portends a weak source of livelihood and inadequate coping mechanism towards disaster risk reduction.

Presumption in urban studies suggests that the longer period of stay in an area by an individual, the deeper the understanding of the environmental condition of the community and the dynamics of different hazardous phenomena in terms of cause, frequency, nature and impacts. The respondent's answer indicated that 18 (6.0%) have stayed in the community within the last 5 years; 30 (10%) have lived in the community between 6 and 10 years; a majority of 154 (51.1%) have resided in the neighbourhood between 11 and 15 years. While 56 (18.6%) and 42 (14.0%) have lived in the neighbourhood between 16 and 20 years, and above 20 years respectively. Given the impression that the respondents ought to be sufficiently aware of the dynamics of flooding, vulnerability of the people and various impacts on hazards on their life and details of coping mechanisms.

Arguably, the result of analysis could be regarded as a strong factor that is needed to build the capacity of the people towards development of a vibrant community with a sense of community necessary for prevention, mitigation, preparedness, preservation of the quality of the environment and consideration of how best to meet the needs of the community in terms of response, rehabilitation and recovery after a disaster.

Income controls and determines the local environmental quality of the area an individual occupies. The analysis shows that 152 (50.6%) respondents earned very low income, 72 (24.0%) could be considered to earn low income, while 60 (20%) fell within the national minimum wage income bracket. Only insignificant 16 (5.3%) of the respondents earned medium income. Implying that the majority of the respondents live below the national poverty level which naturally increased their vulnerability by default and reduced their coping mechanism against future risks.

Analysis of household size indicates that 198 (66.0%) of the respondents is made up of moderately large house sizes of between 7 and 9 persons. Household size 3–6 constitutes 14 (4.6%). Homes with 1–2 sizes constitute 18 (6.0%) of the respondents and 50 (16.6%) and 20 (6.6%) represent household sizes of 10–13 persons and 14 persons respectively. It can be safely concluded that the neighbourhood is experiencing a high-density spatial distribution population which is one of the human factors that can increase people's vulnerability to disasters.

1.7.2 Flooding Occurrences, Causes, Impacts, Coping Mechanism and Effectiveness of Flood Risk Management

Urban flood depicts any overland flow of water over urban streets sufficient to cause nuisance, hazards and disaster. When the respondents were asked questions on the occurrences, experiences and causes of flooding. The respondent's answers to the question of flooding experience indicated a variation in the answer that ranged from 100 (33.3%) any time it rains, suggesting very often; 155 (51.7%) often; 28 (9.3%) not often and 17 (5.6%) have not at all respectively. The implication of the answers given by the significant percentage of respondents confirm that the incidence of flooding is a hazard that challenges the community annually.

On the experience of flooding occurrence, respondents claimed an extremely high level 52 (17.3%), 71 (23.0%) moderately high level and 93 (31.0%) high with the attendant challenges. While 50 (16.6%) and 34 (11.3%) of the respondents claimed low and extremely low level of occurrence respectively. Results of the analysis coupled with weak coping capacity and high levels of vulnerability have continually put many lives and properties at risk.

The responses of the residents to the causes of flooding based on their experience showed a variation that ranged from all the factors that hinged on the respondent's level of education; duration of stay in the neighborhood and their perception. Respondents attributed the causes to frequent rainfall; 42 (14.0%) poor land use planning; 72 (24.0%) dumping of waste in drainage facilities and 30 (10%) high volume of rainfall. Furthermore, 29 (9.6%) claimed hard surfaces; while 21 (7.0%) and 70 (23.3%)

indicated land reclamation and inadequate drainage facilities respectively. The result confirmed that there are a number of factors that induced flooding hazards within the urban planning consideration, while implying that various strategies are also required to control the hazard with a view to creating a resilient urban community. The ensuing risk of climate change that induced flooding devastating effects on the low-income group gives the rationale to examine the impact because of their natural location in the coastal area of Lagos with unique spatial characteristics. Table 1.1 indicates the respondent's claim of impact experience on their life.

Impacts assessment could be felt differently at the level of individual economic livelihood with various implications in terms of preparedness and post-disaster well-being and urban planning recovery [77–80]. The respondents rating of impact dimensions depicted on Table 1.1 using Likert Scale showed a value mean score of above 3.0 for all dimensions, with the implication that the respondents experienced high negative impacts of flooding hazard on various aspects of their life. The high population of the community and the resident's low-income level are factors of the characteristics and circumstances that make the people susceptible to the adverse effect of

Table 1.1 Impact of flooding on the respondents responde

S/N	Impact dimension	Very low negatively	Low negatively	Moderately high negatively	High negatively	Very high negatively	Total	Mean sore
1	Loss of property	29 (9.6%)	39 (13%)	98 (32.7%)	84 (28.0%)	50 (16.7%)	1233	4.11
2	Damage to infrastructure life	26 (8.0%)	30 (10%)	120 (40.0%)	84 (28.0%)	40 (13.3%)	1041	3.47
3	Disruption of economic life	32 (10.6%)	49 (16.3%)	107 (35.7%)	84 (28.0%)	28 (9.3%)	927	3.09
4	Loss of human life	43 (14.3%)	48 (16%)	102 (34.0%)	75 (25.0%)	32 (10.7%)	904	3.02
5	Damage to building	33 (11%)	53 (17.6%)	144 (38.0%)	52 (17.3%)	48 (16.0%)	929	3.09
6	Disruption of family ties	40 (13.3%)	45 (15%)	95 (31.6%)	65 (21.1%)	55 (18.3%)	950	3.20
7	Mobility	32 (10.7%)	51 (17%)	103 (34.4%)	72 (24.0%)	42 (14.0%)	1001	3.30
8	Water supply pollution	30 (10.0%)	55 (18.3%)	123 (41.1%)	53 (17.6%)	39 (13%)	925	3.1
9	Disease's outbreak	36 (12.0%)	48 (16.0%)	111 (37.0%)	45 (15.0%)	61 (20.3%)	950	3.2
10	Depreciation of property value	40 (13.3%)	52 (17.3%)	110 (36.7%)	44 (14.7%)	54 (17%)	920	3.06

Source Author's Field work (2021)

the flooding hazard aggravated by the inadequacy of physical infrastructure in the community.

An understanding and analysis of coping mechanisms which represent different strategies put in place based on indigenous knowledge, available resources, culture and technology to manage flooding hazard at various levels with a view to reducing and mitigating impacts and enhancing level of resilience was investigated. The result at the household level indicated that 50 (16.7%) raised the building's door steps; 39 (13.0%) adapted construction of outlet pipe from the building; while 36 (12.0%) used logs of woods, sand bags and stones and 50 (16.7%) adapted evacuation of water out of the building premises. Other methods used by the respondents are stay indoor strategy 35 (11.7%); building of walls and drainage facilities were carried out by 33 (11.0%); temporary migration value of 27 (9.0%) and redirection of drainage was done by 30 (10.0%).

Coping strategy at the community level usually complement the household strategy and related with preparedness and mitigation measures, management of exceptional circumstance with a view to bringing about a normalcy in life and livelihoods after a disaster. The community level coping mechanism indicated the percentage of the strategies adapted in order of importance by the community. The clearing of drainage facilities with a value of 90 (30.0%); being the most important and the planting of trees constituted 35 (5%) recorded as the least important strategy. Result of the analysis synchronized with Danso and Addo [81] and Mondal's et al. [82] opinion that households employ mixed methods as coping strategies when their communities are at risk from flooding, while Morshed [83] remark that community in the disaster-prone area tend to depend on indigenous knowledge to overcome the threat posed by extreme climate such as flood.

However, in view of the fact that the flooding hazards may subsist following the growing evidence that a global degradation of the environment will cause climate change, coupled with the limited capacity to mitigate the impact at national and local levels, government as stakeholder is increasingly concerned about the threat of flooding to communal safety.

Thus, an assessment of intervention efforts of the government in providing measures to address flooding was considered necessary for investigation from the plurality of strategies perspectives towards flood prevention and a plurality of actor's involvement in mitigating the impact of flooding and post-disaster recovery. The government efforts which aimed at flooding disaster reduction involved desilting the primary and secondary drainages; prevention of indiscriminate dumping; construction of primary and secondary drainages and demolition of illegal structures with a value of 102 (34.0%); 52 (17.3%); 20 (6.7%) and 10 (3.3%) for the efforts respectively. Implying that government intervention cut across different segments of the study area and exceeded other forms of intervention. The principle that confirmed Slavikova [84] opinion that governments do play the most important roles and bears the ultimate responsibility for disaster crisis management in most communities of the developing countries. A further analysis of the effectiveness of government efforts by the respondents revealed the claim by 174 (58%) of the efforts being ineffective, while 126 (42%) claim the efforts were very effective.

The assessment conformed with Oladokun and Proverb [38] assertion of the disconnection between flood risk management system and the urbanization process especially in developing countries where the challenge of the limited interagency coordination in urban development effort is germane towards the creation of a resilient community. The lapses clearly called for an adoption of an alternative planning approach that emphasized the integration between land and flooding management leveraging on structural and non-structural measures with recourse to urban resilience theory to floods [82, 85].

1.7.3 COVID-19 Disaster Impact Assessment and Effectiveness of Management Measures

Previous discussions on the relationship between community residential quality and related infrastructure has been a fundamental crucial element of crisis resilience management both in response to COVID-19 disease and flooding as a climate induced disaster [5]. This section however exposes the discussion and findings of data on assessment of community residential infrastructural facility quality by residents, impact of COVID-19 pandemic on residents and effectiveness of COVID-19 management strategies. This is with a view to charting transformative actions that can be used for post-COVID 19 disaster recovery, sustainable urban planning and community resilience.

Housing quality and its assessment constitute a major area of focus in residential neighbourhood resilience discourse. Studies based on the understanding that the right to adequate housing and access to resilient housing for residents in crowded residential areas will mitigate vulnerability to the floods and the COVID-19 pandemic is scanty. Thus, policy formulation towards enhancing the resident's well-being, is guided by assessing nine housing quality variables using Likert scale. The housing quality assessment indicates that the mean score of the 9 housing quality variables recorded an average mean score below 3.0 The worst housing quality attribute that the residents are dissatisfied with is the sanitation depicting a mean score of 1.63, closely followed by drainage 1.66; 2.01 for water supply and ventilation 2.01. While mean value score of 2.26 is reported for sewage disposal, mean value scores of 2.27; 2.30; 2.43, 2.46 and 2.53 are recorded respectively for waste disposal; illumination; size of the house; bed room size and living room size. These housing quality standard features, infrastructure and services variables that are assessed are vulnerability factors for poor people lacking access to them with negative impacts on their well-being. The results confirmed Tinson and Clair [86] findings that poor communities lacking in infrastructure and services aggravated by poor housing standard features, overcrowding and density are associated with greater spread of COVID-19 disease.

The assessment of the impact of COVID-19 disease carried out utilized Likert scale. The result from the analysis revealed that all the impact dimensions rate is very negative. The result of the analysis of the mean score assessment for all impact dimensions is below 3.0. The worst negative impact with a mean score of 1.76 is recorded for health conditions, which is closely followed by government responses 1.89. Higher mean score values of 2.28; 2.50 and 2.94 are recorded for economic livelihood; mobility and personal relationship respectively.

Additionally, the assessment of the effectiveness of government strategy as response towards the pandemic control demonstrated in the form of strict lockdown; use of face masks; social distancing; water sanitation and hygiene (WASH); isolation; and relief package assistance was carried out. The result of the assessed mean score value of each of the strategies put in place by the government depicted that all the strategies reflected near effective value, moreover, the average mean score for all strategies was below 3.0 the cut off value for effective control. The mean score of 1.97; 2.23; 2.38; 2.76 and 2.93 was recorded for washing of hands with running water, soap and sanitizer (WASH), social distancing; use of face mask; self-isolation and restriction of movement; respectively. This finding correlated with Lawanson [87], Sodiya and Uwala [23], Amzat et al. [88], Oleribe et al. [89] assertion on the inadequacies of government control measures that are put in place to manage the disaster.

The result reflected the imbalance between neighbourhood environmental quality and the urban dweller's well-being. It further justified [41, 80, 88, 90] postulation that COVID-19 pandemic disaster had a significant negative influence on the life of the people at the bottom of the socio-economic spectrum; consequently emotively compelling a local response which has relevance in how urban governance resilience can be used to provide knowledge for stakeholders in the prevention and control of pandemic and develop better informed public crisis prevention and control.

1.8 Conclusions

Flooding and COVID-19 pandemic are hazards that cannot be prevented completely. The hazards occurred from globally changing conditions of rapid population growth and migrations, climatic change and variability and land use change. The coincidence of the disaster necessitated international, national and local attention and concern. Uncontrolled urbanization has the tendency to aggravate these most pressing contemporary global challenges in developing countries. Nigeria's response to the coincidence of COVID-19 pandemic and flood control like other countries have been top priority in Lagos. The area under investigation is an overcrowded and underserved community with basic infrastructure facilities portraying a condition that can continue to provide a ready channel for resurgence of the disasters in the face of lack of coping mechanisms for preparedness, prevention, mitigation, response and recovery.

The current practice towards control of the multiple hazards have been reactive and disintegrated in approach with focus on one hazard at time. The practice is limited in scope because of its inability to consider in a simultaneous manner, the physical, social, environmental and economic consequences of hazardous conditions holistically. Indeed, the practice may be regarded as an insufficient tool for addressing challenges of multi-hazard management.

Consequently, the need for a paradigm shift and new systemic approach becomes necessary as a process that will review, modify the action plans, techniques and protocols in a manner that ensures appropriateness, adequacy, comprehensiveness and support for planning real-time disaster management in a community. The need for critical reflections on the importance of context specific integrated approaches to enhancing the well-being of residents through the implementation of effective planning, response, recovery and adaptation actions in the context of neighbourhood environmental quality improvement is emphasized.

Induced flood hazard from climate change is clearly non-statistic. It varies with the growing evidence that a global degradation of the environment will cause further long-term climate change, extreme weather in the world and the emergence of COVID-19 disease as an indispensable neighbourhood residential quality issue. This requires a reflection on the current practice of urban planning and the making of communities and for the generation of different ways in which communities that are needed are created. The paper concludes that the road map towards creating sustainable communities for an effective management of complex hazard problems is to embrace the concept of urban resilience.

1.9 Recommendations

New approach that is needed and recommended in the management of multi-hazards of flooding and COVID-19 pandemic is urban resilience. Contextually, it is a process approach on how to create future resilient communities that are wanted from the present state of existing communities. Its recommendation hinged on the fact that the contents of urban resilience can be employed as input into strategic and operational planning practice and policy learning in sustainable urban development drive at any scale of urban planning and development.

This recommendation rested on the quest to achieve Sustainable Development Goal (SDG) 11 that focused on making cities inclusive, safe, resilient and sustainable in an urbanizing world. Moreover, urban development in developing countries like Nigeria has focused on projects of regeneration and redevelopment of existing cities which are being challenged by resource constraints. The adoption of new urban resilience concepts and guiding principles offer an option that illustrates ways by which resilient communities of the future that are wanted from the existing ones are restored and created.

In this respect, the four dimensions of urban system that are recommended as foci for the achievement of cities resilience transcend viz, enhancement of communities residents capacities to recover, thrive and innovate; landscaping of the community that involves supply and enjoyment of ecosystem and biodiversity services; building of community infrastructure's structures and services; and institutional arrangement and structure that epitomizes good governance reflecting inclusiveness, partnerships, rules and laws and behavioral change.

Thus, as a way out of escaping from the disasters of the present situation of flooding and diseases in the community, affected by the problem, ability of the community's residents to self-organization and mobilization towards skills and abilities building and capacities to source for new opportunities and create new forms of innovation to act in solidarity in the aftermath of a shock is being recommended. In this regard, the premise on which people's resilience in the community is built rests on developing a sense of community ties, collective self-esteem and cultural identity with a view to enhancing the ability of the community members to absorb shocks, promoting the capacity of people to act with solidarity after a disaster. Emphasis is on the development of a sense of community ties towards the mitigation of a disaster and recovery from it after a shock using a bottom up approach.

The perennial urban flooding caused by climate change and the lingering of COVID-19 pandemic have resulted in the need to examine and understand the community landscape context as an important leverage upon which the affected community can recover and thrive from flooding and diseases. Emphasis on spatial planning strategy is recommended towards the control of transmission of disease and flooding. In this sense, the community need to focus on the recognition of the biodiversity that is available in the neighborhood which contribute to its environmental quality and provided multiple ecosystem services that contribute to well-being and quality of life of residents while taking cognizance of global connectivity and resource imprint of the city that influenced the ability of the community to improve resilience and sustainability. The thrust is on enhancing the community resident's capacities to take up initiatives for restoration of green infrastructure in the community which tend to act in synchrony with the city's master plans to increase permeable surfaces in the urban area thereby improving the storm water absorption and retention in urban spaces.

The aspect of infrastructure resilience is recommended towards improving the community resident's well-being. Infrastructures are the hardware of the communities' needs which enhance the well-being of residents when they are adequately provided, functionally efficient and available to all community inhabitants. With the nature of high density of population living in the community, it is recommended that future investments in infrastructure at the community level should aim at improving the environmental performance and emphasizing the creation of businesses that is based on coupling the creation of business activities and infrastructural facilities that are functional and adequate. The recommendation is anchored on the use of a window of opportunities offered by revitalization strategy and retrofitting of infrastructure as a means of contributing further to community's health and resilience.

The quality characteristics of how the infrastructure can result in delivering community resilience depend on its robustness and adaptability. Thus the community investment on infrastructure must guarantee durability, capacity to provide services that relate to social demands of today and needs of the future over a long period of time irrespective of the magnitude of shock. A pragmatic demonstration of the strategy rest on granting of soft loan in a sustainable manner to the residents with the understanding of seeing housing as a health insurance infrastructure that will result in better buildings that guarantee sanitary condition improvement and comfortable and healthier homes base on the 'new normal situation' concept in such a way that changes the notion of home into wider spectrum that creates the consciousness for people about the likelihood of spending more time inside for the purpose of mitigation.

Additionally, the recommendation is a call for more functional and sustainable urban planning and design that hinged on adoption of green infrastructure strategy to create a healthier environment, by controlling urban flooding using a cost-effective resilient approach that focuses on enhancing the capacity of the community using natural processes for managing flooding impacts. It is understood as a network providing the ingredients for solving urban and climate challenges by building with nature. Such ingredients include proper landscaping that transcend creation of community parks, and street trees with a view to replacing hard paving with permeable and vegetated surfaces towards decreasing community surface runoff.

Embracing the principles of good governance at the local government level is recommended as a vital strategy that facilitates the creation of a resilient community. Recognition and empowerment of Community Development Association (CDA) under the direct supervision of local government is desirable as a vehicle for bringing about the building of a resilient community. The leadership structure and institutional arrangement at this level, must involve partnership between different social actors that include public, private and civil society that are knowledgeable about anticipation of disasters, having an understanding of long standing vulnerabilities experience of the community over a period of time. This is in addition to building the abilities of residents to revitalize community economies and dealing with disaster effectively using bottom-up initiatives to enrich community resilience, by ensuring adequate preparedness for and mitigation of disaster risk and pandemic reduction through adequate planning. Other measures such as establishment of emergency operation centers, continuous assessment of needs and identification of resources for planning for response and implementation of the responses and preparation for community recovery are paramount at the local level of governance.

Acknowledgements The author wishes to express profound appreciation to Victor Omolabi, Goden Uwem and David Shobiye.

References

1. Tacoli C, McGranahan G, Satterthwaite D (2014) Urbanization, rural–urban migration and urban poverty. In: World migration report. International Institute for Environment and Development, London
2. UN DESA (2018) Release of the world urbanization prospect: the 2018 revision. United Nations, Department of Economic and Social Affairs News, 16 May 2018, New York. https://www.un.org/development/des/en/news/population/2018revision-of-world-urbanisation-prospect.html. Accessed 17 March 2021

3. Aremu O (2019) Urbanization and policing in Nigeria rethinking security for national stability. In: Albert IO, Lawanson T (eds) Urban crises and management in Africa, pp 671–685
4. Farrel K (2018) An inquiry into the nature and causes of Nigeria's rapid urban transition. In: Urban forum, vol 29, pp 277–298
5. McKernan L, Weichelt CL (2020) Local struggles for housing rights in the context of climate change, urbanization and environmental degradation. The global initiative for economic, social and cultural rights
6. Simonovic SP, Kundzewicz ZW, Wright N (2021) Floods and the COVID-19 pandemic—a new double hazard problem. Wires Water e1509. https://doi.org/10.1002/wat2.1509. Accessed 04 April 2022
7. Zinda J, Kay DL, Williams LB (2022) Different hazards, different responses: assessment of flooding and COVID-19 risks among upstate New York residents. Socius: Sociol Res Dyn World 8:1–20
8. Izumi T, Das S, Abe M, Shaw R (2022) Managing compound hazards: impact of COVID-19 and cases of adaptive governance during the 2020 Kumamoto flood in Japan. Int J Environ Res Public Health 19(2022):1–16
9. Phillips CA, Caldas A, Carlso C (2020) Compound climate risks in the COVID-19 pandemic. Nat Clim Chang 10(2020):586–588
10. Walton D, Arrighi J, Alast MV, Claudet M (2021) The compound impact of extreme weather events and COVID-19. A first look at the number of people affected by intersecting disasters. IFRC, Geneva
11. Ramalho CE, Hobbs RJ (2012) Time for change: dynamic urban ecology. Trends Ecol Evol 27(3):179–188
12. Frantzeskaki N (2016) Urban resilience: a concept for co-creating cities of the future. Resilient Europe, European Union
13. Seidlein LV, Alabaster GD, Deen J, Knudsen J (2021) Crowding has consequences. Prevention and management of COVID-19 in informal urban settlements. Build Environ 188:1–9
14. Proag V (2014) The concept of vulnerability and resilience. Procedia Econ Financ 18(2014):369–526
15. Wahab B, Atebijie N, Yunusa I (2013) Disaster management in Nigeria rural and urban settlements. Artsmosfare Prints Ibadan, Nigeria
16. Ajibade I, Kerr RB, Mc Bean G (2013) Urban flooding in Lagos, Nigeria: patterns of vulnerably and resilience among women. Glob Environ Chang. https://doi.org/10.1016/j.gloenvcha.2013.08.009. Accessed 26 March 2021
17. Dixon T (2020, June 08) What impacts are emerging from COVID-19 for urban futures? In: Centre for evidence—service to support the COVID-19 response. University of Reading
18. Ezegwu C (2014) Climate change in Nigeria. The impacts and adaptation strategies. https://dx.doc.org/102139/ssrn.2543950. Accessed 21 March 2021
19. Durrizo K, Gunter I (2021) Managing the COVID-19 pandemic in poor urban neighbourhoods: the case of Accra and Johannesburg. World Dev 137:105175
20. Mcfarlane C (2020) The urban poor have been hit hard by Coronavirus. We must ask who cities are designed to serve. The conversation, June 3, 2020. https://theconversation.com. Accessed 11 April 2022
21. Croitoru L, Miranda JJ, Khattabi A, Lee JJ (2020) The cost of coastal zone degradation in Nigeria: cross river, delta and Lagos states. World Bank Group
22. Ohaeri VI, Obinyan A (2020) Spaces for change: Nigeria. In: Weichelt C-L, Mckernan L (eds) Local struggles for housing rights in the context of climate change, urbanization and environmental degradation. Global Initiative for Economic, Social and Cultural Rights
23. Sodiya OO, Uwala VA (2020) Implications of COVID-19 pandemic on high density residential areas of Lagos State. https://www.ResearchGate.net/publication/346007690. Accessed 24 April 2022
24. Balbus J, Crimmins JL, Gamble DR, Easterling KE, Kunkel SS, Sarofim MC (2016) The impact of climate change on human health in the United States: a scientific assessment. U.S. Global Change Research Program, Washington D.C. https://doi.org/10.7930/JOVX0DFW

25. IPCC (2022) Climate change: a threat to human wellbeing and health of the planet. Taking action now can secure our future [EN/AR/RU/ZH]. https://reliefweb.int. Accessed 12 April 2022
26. Ilevbare FM (2019) Investigating effects of climate change on health risk in Nigeria. In: Uher I (ed) Environmental factors affecting human health. Intech Open
27. Anyadike RNC (2009) Climate and change and sustainable development in Nigeria conceptual and empirical issues. Enugu forum policy paper 10. African Institute for Applied Economic Nigeria
28. Royal Society Report (2020) Climate change evidence and causes: update 2020. An overview from the Royal Society and the US National Academy of Sciences
29. Pringle P (2018) Effects of climate change on 1.5 °C temperature rise relevant to Pacific Islands. In: Pacific marine climate report card science review 2018, pp 189–200
30. Singh BJ, Singh O (2012) Study of impacts of global warming on climate change: rise in sea level and disaster frequency. Intech Open. https://doi.org/10.5772/50464. https://wwwintechopen.combooks/global-warming-impacts-and-future-perspective/study-of-impacts-of-global-warming-on-climate-change-rise-in-sea-level-and-disaster-frequency. Accessed 20 April 2021
31. Afolabi O (2020) Is climate change real in Nigeria
32. IPCC (2013) Summary, for policymakers in managing the risks of extreme events and disasters to advance climate change adaptation. In: A special report of working groups 1 and 11 of the intergovernmental panel on climate change. Cambridge University Press, Cambridge, UK; New York, pp 1–19
33. Olaniyi OA, Ojekunle ZO, Anyo TT (2013) Review of climate change and its effects in Nigeria ecosystem. Int Journey Afr Asia Stud 1(2013):57–64
34. Adenekan IO (2016) Flood risk management in the coastal city of Lagos, Nigeria. J Flood Risk Manag 255–264
35. Atufu CE, Holt CP (2018) Evaluating the impacts of flooding on the residents of Lagos, Nigeria. Urban Water Syst Floods II 184:81–90
36. Few R (2003) Flooding, vulnerability, and coping strategies: local response to a global threat. Prog Dev Stud 3(1):43–58
37. Nasiri H, ShahoMohammadi S (2013) Flood vulnerability index a knowledge base for flood risk assessment in urban areas. J Nov Appl Sci 203(2):269–272
38. Oladokun VO, Proverb D (2016) Flood risk management: a review of the challenges and opportunities. Int J Saf Secur Eng 6(3):485–497
39. Farha L, Perucca J (2020) Housing and climate crisis. In: Weichelt CL, McKernan L (ed) Local struggles for housing rights in the context of climate change urbanization and environmental degradation. Misereor Ihr Hilfswerk
40. Dos Santos WG (2020) Natural history of COVID-19 and current knowledge on treatment therapeutic option. https://doi.org/10.1016/biopha.2020.11093. http://www.ncbi.nlm.nihi.gov/pmc/article. Accessed 30 March 2021
41. Connolly C, Ali SH, Keil R (2020) On the relationship between COVID-19 and extended urbanization. Dialogues Hum Geogr 10(20):213–216
42. Baker E, Lester LH, Bentley R, Beer AP (2016) Poor housing quality: prevalence and health effects. J Prev Interv Community 44(4):219–232
43. Mouratidis K (2020) Neighbourhood characteristics, neighbourhood satisfaction and well-being: the links with neighbourhood deprivation. Land Use Policy 99:104886
44. Omolabi AO (2018) Residential satisfaction and revitalisation of public low-cost housing residential estate in Lagos, Nigeria. Unpublished (PhD) thesis, University of Kwazulu, Natal, Durban, South Africa
45. Dwijendra NKA (2013) Quality of affordable housing project by public and private developers in Indonesia. The case of Sarbagita Metropolitan Bali. J Geogr Reg Plan 6(3):69–81
46. Lee S, Parrott KR, Aln M (2014) Housing adequacy. A well-being indicators for elderly households in Southern US communities. Fam Consum Sci Res J 42(3):235–251

47. Donaldson C, Rolfe S, Garnham L, Godwin J, Anderson I, Seaman P (2020) Housing as a social determinant of health and well-being. Developing an empirically-informed realist theoretical framework. BMC Public Health 1138
48. Gilbertson JG, Gormandy D, Thompson H (2008) Good housing and good health? A review and recommendations for housing and health practitioners. A sector study UK housing corporation. https://www.health_housing_20060816144328.pdf
49. Ulmer JM, Wolf K, Desiree RB, Lawrence DF (2016) Multiple health benefits of urban tree canopy: the mounting evidence of a green prescription. Health Place 42:54–62
50. UNISDR (2012) Making cities resilient report 2012. United Nations International Strategy for Disaster Reduction Secretariat
51. Meerow S, Joshua P, Newell MS (2016) Defining urban resilience: a review. Landsc Urban Plan 47(2016):38–49
52. Prasad N, Rangheiri F, Shah F, Trohanis Z, Kessler E, Sinha R. (2009) Climate resilient cities: a primer on reducing vulnerabilities to disasters. The World Bank, Washington
53. Cutter SL, Barnes L, Berry M, Burto C, Evans E, Tate E, Webb J (2008) A place—based model for understanding community resilience to natural disasters. Glob Environ Chang 18:598–606
54. Willner SN, Levermann A, Zhao F, Frieler F (2018) Adaptation required to preserve future high-level river flood risk at present levels. Sci Adv 4(1):1–8. https://doi.org/10.1126/sciadv.aao1914
55. Garciar-Mira R, Uzzell D, Martínez JR, Deus JE (2005) Housing, space and quality of life. Ashgate Publishing Limited
56. Pacione M (2003) Urban environment quality and human well-being. A social geographical perspective. J Landsc Urban Plan 65(2003):19–30
57. Prochorskaite A, Maliene V (2003) Health, well-being and sustainable housing. Int J Strateg Prop Manag 17(1):44–57
58. Balestra C, Sultan J (2013) Home, sweet home. The determination of residential satisfaction organization economic co-operation and development. STD. DOC 5 working paper 54
59. Garrett TA (2010) Economic effects of the 1918 influenza pandemic: implications for a modern-day pandemic. Working paper CA0721, 2007. Federal Reserve Bank of St. Louis. https://www.stlouisfed.org/~/media/files/pdfs/communitydevelopment/researchreports/pandemic_flu_report.pdf
60. Glasser EL (2011) Cities, productivity, and quality of life. Science 333(6042):592–594. https://doi.org/10.1126/science.1209264
61. Chowell G, Bettencourt LM, Johnson N, Alonso WJ, Viboud C (2008) The 1918–1919 influenza pandemic in England and Wales: spatial patterns in transmissibility and mortality impact. Proc Biol Sci 275(1634):501–509
62. Hamidi S, Ewing R, Sabouri S (2020) Longitudinal analyses of the relationship between development density and the COVID-19 morbidity and mortality rates: early evidence from 1,165 metropolitan counties in the United States. Health Place 64(2020):102378. https://doi.org/10.1016/j.healthplace.2020.102378
63. Bond L, Kearns A, Whitely E (2012) Exploring the relationships between housing, neighborhood and mental well-being for residents of deprived areas. BMC Public health 12(48)
64. Opoko AP, Oluwatayo AA, Ezema IC, Opoko CA (2016) Resident's perception of housing quality in an informal settlement. Int J Appl Eng 11(4):2525–2534
65. Ma LILI (2018) A brief analysis of the relationships between housing, mental health and well-being under the eco-city context. Adv Econ Bus Manag Res (AEBMR) 60(2018):930–935
66. Hoffman AV (2008) The lost history of urban renewal. J Urban 1(3):281–301
67. Holding E, Blank L, Crowder E, Goyer E (2020) Exploring the relationship between housing concerns, mental health and well-being: a qualitative study of social housing tenets
68. Lee GK, Cham EH (2018) The analytic hierarchy process (AHP) approaches for assessment of urban renewal proposals. Soc Indic Res 89(1):155–168
69. Vidgor JL (2010) Is urban decay bad? Is urban revitalization bad too? J Urban Econ 68(2011):277–289

70. McKenna A (2021) Mushin-Nigeria: encyclopedia Britannia, 20 September, 2010. https://www.hritannica.com/place/mushin. Accessed 31 March 2021
71. Roberts AA, Balogun MR, Sekoni AO, Inem VA, Ochukoya OO (2015) Health seeking preferences of residents of Mushin LGA, Lagos: a survey of preferences for provision of maternal and child health services. J Clin Sci 12(1):9–13
72. Cross JE (2001, November 2–4) What is sense of place? In: Paper prepared for the 12th headwaters conference. Western State College
73. Akani O (2020) COVID-19: Lagos LGAs and number of confirmed cases. https://www.vanguardngr.com. Accessed 20 April 2021
74. Harwell MR (2012) Research design in qualitative, quantitative and mixed methods section III. In: Opportunities and challenges in designing and conducting inquiry. University of Minnesota
75. Isreal GD (2009) Determining sample size. University of Florida IFAS Extension Publication NPE OD6. https://edis.ifas.ufi.edu./pd006. Accessed 20 April 2022
76. Canny G (2006) 5 point versus 6 point Likert Scales. http://www.geocanywrytestuff.com/swa69773.htm. Accessed 30 April 2021
77. Duggal R (2020) Mumbai's struggles with public health crisis from Plague to COVID-19. Econ Polit Wkly 55(21):17–20
78. Enarson E, Fothergill A, Peek L (2006) Gender and disaster, foundations and directions. In: Rodriguez H, Quarantelli EL, Dynes RR (eds) Handbook of disaster research. Springer, New York
79. Sultana F (2010) Living in hazardous waterscapes; gendered vulnerabilities and experiences of floods and disasters. Environ Hazards 9(1):43–53
80. Wade L (2020) An unequal blow. Science 368(6492):700–703
81. Danso SY, Addo IY (2016) Coping strategies of households affected by flooding: a case study of Sekondi—Takoradi Metropolis. Urban Water J 14(5):539–545
82. Mondal HS, Murayama T, Nishikizawa S (2021) Determinants of household-level coping strategies and recoveries from riverine flood disasters. Empirical evidence from the right bank of Teesta River Bangladesh. Climate 9(4). https://doi.org/10.3390/cli9010004
83. Morshed M (2007) Indigenous coping mechanism in combating flood. Unpublished MSc thesis in Disaster Management BRAC University Dhaka, Bangladesh
84. Slavikova L (2018) Effects of government flood expenditures: the problem of crowding-out. J Flood Risk Manag 11(2018):95–104
85. Liao KH (2021) A theory on urban resilience to floods—a basis for alternative planning practices. Ecol Soc 17(4):48
86. Tinson A, Clair A (2020) Better housing is crucial for our health and the COVID recovery health. www.health.org.uk/publications/hug-reads/betterhousingiscrucial-for-our-health-and-the-COVID-19recovery16/4/2021
87. Lawanson T (2020) Lagos size and slum will make stopping the spread of COVID-19 a tough task. The conversion April 1, 2020
88. Amzat J, Aminu K, Danjibo MC (2020) Coronavirus outbreak in Nigeria. Burden and socio-medical response during the first 100 days. Int J Infect Dis 98:218–224
89. Oleribe O, Ezechi O, Osita-Oleribe P (2020) Public perception of COVID-19 management an response in Nigeria: a cross sectional survey. BMJ Open 10:e041936. https://doi.org/10.1136/bmjopen-2020-041936
90. Wasdani KP, Prasad A (2020) The impossibility of social distancing among the urban poor. The case of an Indian slum in times of COVID-19. Local Environ 25(5):414–418

Chapter 2
Beyond COVID-19: Planning the Mobility and Cities Following "15-Minute City" Paradigm

Tiziana Campisi and Kh Md Nahiduzzaman

2.1 Introduction

The pandemic period has profoundly changed travel patterns and use urban amenities with deserted roads, short journeys and numerous restrictions. Global mobility has been negatively affected by the pandemic. Public [1] and shared urban transport systems [2, 3] have been critically suffered. Several studies are published at local and national levels to analyse the differentiated travel habits and mode choice behaviour by the users in pre- and post-pandemic periods [4, 5]. Having an optimistic view, while we are planning for a post-pandemic normalcy with an evolving urban structure, strategic direction for mobility must not (re-)emerge from the recent pandemic only. It needs to cope with 2.3 billion additional people who are expected to live in urban areas by 2050, the increasing demand for per capita transport, a resilient post-pandemic response and an urgent need to reduce greenhouse gas emissions [6, 7].

People have the urge and need to access public spaces. However, the fear of using public spaces while complying with the changing restrictions has not only decreased the level of utility, but also created post-pandemic panic that may continue to hinder the embedded amenities. Thus, the contemporary planning principles dictating the design and efficacy in the use of public recreational spaces are facing a sheer challenge of what guidelines would reinstate the intended use while dealing with the post-pandemic fear [8]. Moreover, lack of access to such urban services could be a major cause of inequality. Providing better urban access to the most vulnerable road users (e.g. elderly, children, disabled, etc.) and connecting them to services and opportunities would be a way to overcome such inequality [9, 10]. As it turns out,

T. Campisi (✉)
Faculty of Engineering and Architecture, University of Enna Kore, 94100 Enna (EN), Italy
e-mail: Tiziana.campisi@unikore.it

K. M. Nahiduzzaman
The University of British Columbia Okanagan, Kelowna, Canada
e-mail: Kh.Nahiduzzaman@ubc.ca

© The Author(s), under exclusive license to Springer Nature Singapore Pte Ltd. 2023
L. Zhang et al. (eds.), *The City in an Era of Cascading Risks*, City Development: Issues and Best Practices, https://doi.org/10.1007/978-981-99-2050-1_2

the twin challenges of post-pandemic fear and unequal access to the public spaces has further confronted the fundamental planning, design and operational guidelines.

The recent pandemic has highlighted the need to downscale the travel volume while enhancing services and technology in every neighbourhood. Therefore, local governments, mobility experts and transport operators have a key role in minimising travel demand by avoiding the need to travel, reducing the length of journeys and creating conditions to make walking, cycling and public transport attractive. Shorter and heightened number of local trips make walking and cycling increasingly suitable modes of transport. More actions will have to be taken to design, build and maintain high quality connections that make walking and cycling safe, easy and direct, and suitable for people with reduced mobility [11]. Indeed, creating and connecting with shared quality public spaces has important social and environmental benefits, in addition to the positive health impacts of the increased volume of walking and cycling [12]. Public spaces can also recover their function to spend time and connect where people can meet, play and recreate themselves.

Through the provision of safer, cheaper, accessible and cleaner means of transport, sustainable urban mobility turns out to disproportionately benefit the vulnerable populations, including children, elderly, poor and people with disabilities [13]. Thus, it is possible to define three essential elements in order to achieve a reduction in widespread transport demand:

- Reducing transport demand through improved planning requires widespread and systematic capacity building at local and national levels to equip all cities with the means to reduce transport demand in favour of sustainable and low-carbon cities [14].
- Promoting sustainable lifestyles and green forms of mobility at an early stage [15].
- Fostering appropriate national policies and planning frameworks to connect green mobility with public spaces to cater the needs for all [16].

The city model that covers the aforementioned elements in recent urban planning practices are those related to the "15-Minute City" or proximity city. After a thorough literature review, this paper aims to critically examine the inherent advantages of the "15-Minute City" for a new urban living in the current and post pandemic era.

2.2 "15-Minute City" and City of Proximity Definition

The "15-Minute City" is a city of living that extends from residence to the neighbourhood and different activities and services that can be found in it. It, therefore, proposes a vision of contemporary living on the new idea of proximity and the values it can bring with it. Thus, it does not correspond to the city of suburbs.

Travel restrictions have also encouraged the virtualization of many activities. On the other hand, the need to find everything while moving as little as possible has been the pivotal idea of "15-Minute City", which revolves around the principle of

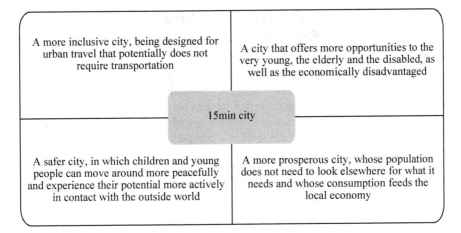

Fig. 2.1 Key advantages of "15-Minute City"

proximity. Opposing to the zoning processes that shaped the twentieth century's Italian suburbs of separating areas with different uses (e.g. residential, commercial, manufacturing, etc.), the "15-Minute City" is proposed as a space within which the inhabitants can find all the urban services and amenities by moving on foot or by bicycling for no more than a quarter of an hour. This model is not only related to sustainable environmental strategies, but also a futuristic idea, which will not make people spend hours on transport, looking for parking or queuing at central offices. It is designed to accommodate a large number of inhabitants in one place. The advantages of such a people-friendly city can be summarised as shown in Fig. 2.1.

Most of the European cities date back to well before the zoning boom and have a distributional network that, due to its spontaneous evolution over the centuries, is rarely divided by the intended use. Thus, the transformation of existing centres into "15-Minute City" would not require a total overhaul of the existing activities. However, some actions aimed at renewing services and activities, limiting what is present in supernumerary and integrating what is scarce or missing. Due to the evolution of new generations of digital tools, such as IoT and Big Data, it is now easier to develop complex analyses for the new challenges that are multi-scalar in nature. Most of the traditional tools are inadequate to measure the indicators of proximity and well-being based on the availability and quality of services, infrastructure, green spaces and mobility related to the location, characteristics and needs of residents. Such transformation is not only critical during COVID-19 pandemic when there is an immediate need for decisions on the land use and urban space to return to "normalcy", but also to assess the long-term opportunities for "meanwhile" use of a space. From a methodological perspective, a need to move away from the "top-down" to "bottom-up" approach based on the specificities of each place and its inhabitants along with their characteristics and needs is evident [16]. Moreover, a radical shift of focus from a traditional governance to new performance-based objectives linked to the expectations of the end users (e.g. citizens, businesses, etc.) is needed [17].

2.3 The Impact of City Space on Well-Being: Importance of Proximity, Harmony and Accessibility

The city places where we live have a great impact on our well-being: the positive result of everyday living is given by the quality of the interior spaces and the outdoor spaces where we move. If the mere fact of moving between the places where we live (home, workplace, public places, etc.) can be stressful, how can we live a healthy everyday life? Danish architect Jan Gehl also commented on this issue back in 2012: studying citizens and their needs is fundamental to any architectural project. The distances we travel every day, the services we have access to near our homes and work, and the ease with which we can move between the fundamental places of our lives are important factors in optimising urban planning and exemplifying life in the urban context [18]. The 20-minute vision of neighbourhoods envisages precisely that all these elements are in balance. According to Mackness and Barrett [19], this model of city is related to 'a concept of the city that goes in the opposite direction of modern urbanism': now more than ever we have the opportunity, if not the urgency, to rethink urban spaces, also to meet new—and unstoppable—changes linked, for example, to the recent need to work from home [20]. There are four characteristics of such places:

- Ecology—to create a green and sustainable city
- Proximity—to live at a reduced distance from one's work activities
- Solidarity—to create links between people
- Participation—to actively involve citizens in the transformation of the spaces they live in.

The implementation of the concept of a "15-Minute City" means that a city is able to guarantee wider social equality and access to the main urban services to citizens and city users. A number of studies have developed a series of indices, which can detect porosity, crossing and attractiveness. The latter is a composite index that can be used to improve pedestrian accessibility in central urban locations [21]. The "15-Minute City" is thus an attempt by the urban environmentalists to reconcile with the human beings [22]. To achieve it, we start from three cornerstones:

- the rhythm of cities should follow human beings, not to serve the cars;
- neighbourhoods should be designed for living, dwelling and thriving without the need for constant displacement;
- moving is, therefore, a choice, not an obligation.

2.4 Proximity Cities Around the World and Takeaways

A study conducted by [23] compared three cases of cities that have adopted the vision of "15-Minute City" and emphasized a shift in planning from neighbourhood accessibility and proximity to urban functions within neighbourhoods, with major systematic changes in resource allocation patterns and governance schemes at the city level. In general, it is possible to build a city on "proximity" if: (i) the model of the "15-Minute City" is the main element of spatial and functional organisation of the unit of neighbourhood unit; (ii) its strategic adoption in the planning doctrine; and (iii) city-wide application of the plan. The parameters of inclusion, health and safety are the pillars of adoption for this model. Table 2.1 summarises the thematic strategies of the European, Australian, Asian and US cities.

The proximity city model has been taken up by several European cities, and Paris was the first European metropolis to adopt sustainable planning through the "Paris en Commun" strategy, i.e. considering urban space based on the concept of proximity, to reduce car travel in the city, favouring travel by bicycle or on foot. During the months of the pandemic, almost every major city redesigned its cycle lanes, widened its spaces for pedestrians and outdoor parking. This strategy has placed it at the centre of every Parisian. All the essential activities making up a large part of everyone's life are within a quarter of an hour: studying, working, shopping, being outdoors, exercising, going to the doctor, going out and having fun. Such an urban model brings social and cultural innovation, as well as a potential response to the pandemic: if there is less space to travel, the risk of mass travel, congestion and contagion is also reduced [36]. These actions have already been applauded by citizens, who in fact see this decision "as one of the emblematic steps of the next decade", and which will soon be accompanied by the introduction of new bicycle lanes, a plan worth 350 million euros, and the elimination of 60 thousand

Table 2.1 City of proximity across the world

Continent	Country	City	City model/strategy	References
Europe	France	Paris	15 min city	[22]
	Britain	n.d.	High streets	[24]
	Spain	Barcelona	Superblocks	[22]
	Denmark	Copenhagen	15 min city	[25]
	Netherlands	n.d.	15 min city	[26]
	Italy	Turin	15 min city	[27]
		Milan		
		Bologna		
USA	Oregon	Portland	20 min districts	[28]
	Texas	Houston	The walkable place	[29]
	New York	New York	15 min city	[30]
	Colombia	Bogotà	15 min city	[31]
Asia	China	Shanghai	20 min city	[32]
	Singapore	n.d.	45 min city	[33]
Australia	Victoria	Melbourne	20 min districts	[34]
	New South Wales	Sydney	City proximity	[35]

parking spaces for private cars, as promised in the election campaign. In Italy, Milan was the first major city to adopt this model for future development. The strategy adopted has been linked to the definition of integrated residential neighbourhoods, even outside the central area, in which housing, offices, factories, public services and green spaces can coexist in order to reduce the phenomenon of work commuting and contribute to the decongestion of public transport and traffic at peak times. Moreover, Milan has left no one behind, as demonstrated by the adoption of the new tariff plan for season tickets for shared mobility services that offer discounts for the low-income passengers. An agreement has also been signed for a taxi sharing service as an alternative to the classic public services, offering those who want or need to travel in total safety. Since its penultimate urban mobility plan (2013), Barcelona has also embraced a concept similar to Moreno's "City of a Quarter of an Hour", designing so-called "Superblocks": mainly pedestrian blocks, with access for only a few authorised vehicles, which represent small communities within the city and are interconnected to other urban blocks by external roads. Thanks to a proactive spatial planning model in the Netherlands that has been in place for many years. Recent studies have shown that more than 80% of the Dutch urban settlements now meet the characteristics of the "15-Minute City". Outside Europe, it is the city of Sydney, in Australia, which for some years has been proud to be a 20-minute city, highlighting how this concept of urban space is leading to an improvement in both the environment and the quality of life of its residents. Similarly, Portland (Oregon) has created 20-minute neighbourhoods, mainly pedestrianised, which are the cornerstone of the city's actions to combat the current climate crisis. In Melbourne, on the other hand, a 30-year programme has been set up, from 2017 to 2050, to move the city towards the dimension of "20-minute neighbourhoods". The pilot project is currently underway, and measurements are also being taken of the exact distances and times needed to access services. The mobility related to the city model analysis involves increasing pedestrian and cycle mobility, followed by public and shared mobility, with the use of private vehicles being the last option. In an urban context, public transport must be integrated with smart and sustainable urban mobility. Only in this way, users would be able to switch from one means of transport to another in a simple, integrated or multimodal way that is economical and compatible with environmental regulations. The transformation of urban mobility cannot wait any longer: with the end of the pandemic, travel patterns have become much more 'individual' (private car or for short distances, motorbike, bicycle or private scooter), a situation that could recreate situations of increasing traffic congestion as shown in Fig. 2.2.

Public transport was hit hard by COVID-19; its use first plummeted, then somewhat recovered, and declined again with the waves of the pandemic. The proximity city model can revive cities after the pandemic phase and at the same time enforce decarbonisation policies. Table 2.2 shows the upward and downward trends of the main variables characterising this type of city.

The first data on the resumption of mobility, after the gradual return to the workplace and the reopening of schools, underline new trends after the lockdown. As several sources point out, safety is at the top of urban commuters' needs, which is why people tend to travel alone, preferring individual means of transport, such as

2 Beyond COVID-19: Planning the Mobility and Cities Following …

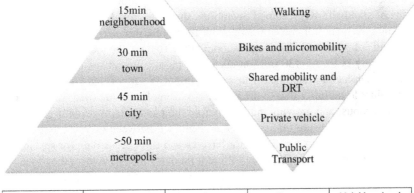

Fig. 2.2 The pyramid approach to the evolution of cities and mobility

Table 2.2 Elements distinguishing the implementation of the 15-Minute City model

	Increasing elements	Decreasing elements
Transport supply (infrastructures and services)	Cycle paths	Roads congested with motor vehicles
	Footpaths	Local public transport (partially)
	Parklets	Shared transport (partially)
	Shared and private micromobility	Reduction of environmental emissions
	Demand responsive mobility	Reduction of weak mobility demand
Social, technological and environmental aspects	Electric mobility and charging areas	Reduction in environmental emissions
	5G and IoT	Areas without a WIFI signal
	Greater aggregation and sociality	Indifference among people living in the same neighbourhood

electric scooters and regular and pedal-assisted bicycles. Public transport, after many months of standstill, has resumed operations showing great limitations in services, reporting a vertical collapse in the demand for public mobility and revenues from tickets and subscriptions dropping by about 80%. To date, the local public transport sector in Italy is losing 130 million euros per month. Incentives to revive the car market, which had suffered heavy losses, seem not to support the concept of green and sustainable mobility due to the increase of vehicles on the road. Even sharing mobility, a robust and growing industry before the pandemic, is revealing its limits, as it is difficult to ensure that vehicles are sanitised between uses. Globally, more than 15 companies active in the sector have closed in 2019. A plan for sustainable urban mobility is needed. There are different models, but everywhere they focus on encouraging the presence of a variety of services and activities necessary for urban life in each area accessible by slow mobility, fifteen minutes on foot or a few minutes by bicycle, in dense, unspecialised areas where different activities coexist and families from different social classes reside. The opposite of gentrification, which has divided cities into ghettos for the rich, the modest and the poor. The proximity city drastically reduces urban pollution and encourages community recreation, which the specialised city and the technological autism of the home has shattered.

Urban development has undergone different transformations over the years. While the European cities are more in favor of a compact structure, cities in North America tend to generously respond to augmenting demand of the population and reflect that in the increasing expansion of their spatial boundaries, making them highly dispersed or sprawl in nature. The geometry of European cities, and in particular the Italian one, has been affected by the concept of small neighbourhoods—with a radius of less than 1 km. In the predominant contexts of Italian urbanization, attempts have been made to concentrate as many activities as possible within a neighbourhood, especially in historical cities that have a radial or concentrated development. Commuting on foot or other non-motorized modes has been a contemporary feature in the Italian urban areas. Moreover, the old cities have the criticality of old and narrow streets while the new neighbourhoods have wide and new streets that are designed to be used for public transit. However, in the US cities, the size of the neighbourhood is larger that follows the suburbanization or sprawl strategy. Often, mobility within and across the neighbourhoods is highly unlikely to be possible on foot or bike as distance is greater than a kilometre. Following the paradox of "15-Minute City", sub-urban and sprawl cities have the potential to be located too close to the urban centres and the expansion should be linked to public transit as well.

However, while the fundamental assumptions of social cohesion for inclusive communities are under question, the post-pandemic fear adds an unorthodox dimension of challenge to building cohesive spaces. On one hand, there is an urging sense of inclusive and resilient communities calling for interactive recreational spaces for all, restriction-posed during COVID-19 and fear in the post-pandemic era stand as systemic impediments for pervasive cohesion, on the other hand [6]. This stems the need for a promising solution that integrates shorter trips and facilitates fearless interaction in the recreational urban spaces that are accessible to every socio-income class across the cities [37]. Moreover, that solution needs to be sensitive in terms of short, medium and long temporal lines. "15-Minute City" turns out to be a well-fit strategy to address this emerging paradox of planning and transportation studies.

2.5 Conclusions

The critical need of the post-pandemic era is about finding sustainable solutions that do not neglect safety and proximity to services. The solution for a new start lies in rethinking travel in a green way. If the future cities have everything, we need within 15–20 min from home, it will be essential to adopt a similar attitude at home. The pandemic that is transforming cities is also affecting our homes: opportunities are arising to redesign everyday spaces to make them multifunctional and more liveable, for example, by furnishing very small balconies or expanding the home spaces into the garden and connecting the community spaces with visible tracks of walking, bicycling, e-scooting, etc. This paper offers a brief critical overview of the spread of the "15-Minute City" model and the evolution of sustainable mobility in the backdrop of the need to reform guidelines for planning, design and operation of urban spaces. However, the model is also suffering from a few drawbacks. It does not take into account the infamous legacy of urban inequity, prevalence of segregated neighbourhoods, amenity inequity and discriminatory policies of the urban public spaces [38, 39]. It remains to be seen whether this city model tends to overcome its inherent disadvantages during the post pandemic era when the need for sustainable and healthy cities and neighbourhoods takes over the segregation and inequity among the different income classes. This paves out a foundation for further research that looks at its applicability against the changing need for urban design and plan to accommodate people from diverse socio-economic backgrounds and shifting need for mobility.

Because of COVID-19, although there is an apparent demand for single unit housing in the urban fringe or remote locations, it is premature to claim that this is going to be a new trend in the future. While global pandemic, such as COVID-19 understandably would not last long, as history suggests, such a trend should be recognized as an immediate response to address the current pressing need [6]. Despite numerous visible benefits, further studies should warrant if "15-Minute City" would be a "new" norm of living in the future cities.

References

1. Campisi T, Basbas S, Al-Rashid MA, Tesoriere G, Georgiadis G (2021) A region-wide survey on emotional and psychological impacts of COVID-19 on public transport choices in Sicily, Italy. Trans Transp Sci 2:1–10
2. Abdullah M, Ali N, Hussain SA, Aslam AB, Javid MA (2021) Measuring changes in travel behavior pattern due to COVID-19 in a developing country: a case study of Pakistan. Transp Policy 108:21–33
3. Torrisi V, Campisi T, Inturri G, Ignaccolo M, Tesoriere G (2021, March) Continue to share? An overview on Italian travel behavior before and after the COVID-19 lockdown. AIP Conf Proc 2343(1):090010 (AIP Publishing LLC)
4. Moslem S, Campisi T, Szmelter-Jarosz A, Duleba S, Nahiduzzaman KM, Tesriere G (2020) Best–worst method for modelling mobility choice after COVID-19: evidence from Italy. Sustainability 12(17):6824
5. Abdullah M, Ali N, Hussain SA, Aslam AB, Javid MA (2021) Measuring changes in travel behavior pattern due to COVID-19 in a developing country: a case study of Pakistan. Transp Policy 108:21–33
6. Nahiduzzaman KM (2021) COVID-19 and change dynamics in the transformational cities. J Urban Manag 10(2):95–96
7. Nahiduzzaman KM, Lai SK (2020) What does the global pandemic COVID-19 teach us? Some reflections. J Urban Manag 9(3):261
8. Shrestha N, Shad M, Ulvi O, Khan M, Karamehic-Muratovic A, Nguyen U, Baghbanzadeh M, Wardrup R, Aghamohammadi N, Cervantes D, Nahiduzzaman KM, Zaki R, Haque U (2020) The impact of COVID-19 on globalization. One Health 100180
9. Mrak I, Campisi T, Tesoriere G, Canale A, Cindrić M (2019, December) The role of urban and social factors in the accessibility of urban areas for people with motor and visual disabilities. AIP Conf Proc 2186(1):160008 (AIP Publishing LLC)
10. Campisi T, Ignaccolo M, Inturri G, Tesoriere G, Torrisi V (2021) Evaluation of walkability and mobility requirements of visually impaired people in urban spaces. Res Transp Bus Manag 40:100592
11. Rahman MT, Nahiduzzaman K (2019) Examining the walking accessibility, willingness, and travel conditions of residents in Saudi cities. Int J Environ Res Public Health 16(4):545
12. Campisi T, Basbas S, Skoufas A, Akgün N, Ticali D, Tesoriere G (2020) The impact of COVID-19 pandemic on the resilience of sustainable mobility in Sicily. Sustainability 12(21):8829
13. Campisi T, Mrak I, Errigo MF, Tesoriere G (2021, March) Participatory planning for better inclusive urbanism: some consideration about infrastructural obstacles for people with different motor abilities. AIP Conf Proc 2343(1):090006 (AIP Publishing LLC)
14. Chen F, Zhu DJ (2009, January) Research on the content, models and strategies of low carbon cities. Urban Plan Forum 4:7–13
15. Pinna F, Masala F, Garau C (2017) Urban policies and mobility trends in Italian smart cities. Sustainability 9(4):494
16. Nahiduzzaman KM, Holland M, Sikder SK, Shaw P, Hewage K, Sadiq R (2021) Urban transformation toward a smart city: an e-commerce—induced path-dependent analysis. J Urban Plan Dev 147(1):04020060
17. Karsten L (2009) From a top-down to a bottom-up urban discourse: (re) constructing the city in a family-inclusive way. J Hous Built Environ 24(3):317–329
18. Friedman A (2021) Sense of place, human scale, and vistas. In: Fundamentals of sustainable urban design. Springer, Cham, pp 75–83
19. Mackness K, White I, Barrett P (2021) Towards the 20 minute city
20. Botzoris GN, Profillidis VA, Galanis AT (2016, March) Teleworking and sustainable transportation in the era of economic crisis. In: 5th international virtual conference on information and telecommunication technologies (ICTIC 2016), Žilina, Slovakia
21. Balletto G, Ladu M, Milesi A, Borruso G (2021) A methodological approach on disused public properties in the 15-minute city perspective. Sustainability 2021(13):593

22. Moreno C, Allam Z, Chabaud D, Gall C, Pratlong F (2021) Introducing the "15-Minute City": sustainability, resilience and place identity in future post-pandemic cities. Smart Cities 4(1):93–111
23. Pozoukidou G, Chatziyiannaki Z (2021) 15-Minute city: decomposing the new urban planning's Eutopia. Sustainability 2021(13):928
24. Hall SM (2011) High street adaptations: ethnicity, independent retail practices, and localism in London's urban margins. Environ Plan A 43(11):2571–2588
25. Byrne J (2021) The 15 minute city: possible or problematic? Planning News 47(4):24–25
26. Bontje S (2021) The advantage of the 15-minute city. https://mobycon.com/updates/the-advantage-of-the-15-minute-city/
27. Rasero F (2020) Città dei 15 minuti, un modello urbano sostenibile basato sulla prossimità. https://www.ehabitat.it/2020/12/07/citta-dei-15-minuti-modello-urbano-sostenibile-prossimita/
28. Marshall B (2020) What is a 20-minute neighbourhood. https://www.domusweb.it/en/news/gallery/2020/09/16/the-20-minute-neighbourhood.html; https://www.houstontx.gov/planning/Commissions/committee_walkable-places.html
29. Kalbfeld JR (2021) Privilege in proximity: neighborhood change and social control in New York City, Doctoral dissertation, New York University
30. Guzman LA, Arellana J, Oviedo D, Aristizábal CAM (2021) COVID-19, activity and mobility patterns in Bogotá. Are we ready for a '15-minute city'? Travel Behaviour and Society 24:245–256
31. Yu Z, Hino Y (2001) A basic study on characteristics of person trip in Shanghai City. Memoirs-Faculty of Engineering, Osaka City University, vol 42, pp 71–76
32. Moser D (2020) Singapore: the 45 minute city, transformative urban mobility initiative (TUMI). https://www.transformative-mobility.org/news/singapore
33. Khor LA, Murray S, Dovey K, Woodcock I, Pasman R (2013, November) New urban territories: spatial assemblies for the 20-minute city. In: 6th state of Australian cities conference, 26–29 November 2013, Sydney, Australia
34. Herath S, Jayasekare AS (2021) City proximity, travel modes and house prices: the three cities in Sydney. J Hous Built Environ 36(2):407–431
35. Moreno C, Allam Z, Chabaud D, Gall C, Pratlong F (2021) Introducing the "15-Minute City": sustainability, resilience and place identity in future post-pandemic cities. Smart Cities 4(1):93–111
36. Nahiduzzaman KM, Aldosary A, Ahmed S, Hewage K, Sadiq R (2020) Urban cohesion vis-a-vis organic spatialization of "Third places" in Saudi Arabia: the need for an alternative planning praxis. Habitat Int 105:102258
37. O'Sullivan F (2021) Where the '15-minute city' falls short'. Bloomberg CityLab 13
38. Nahiduzzaman KM, Aldosary AS, Mohammed I (2019) Framework analysis of E-commerce induced shift in the spatial structure of a city. J Urban Plan Dev 145(3):04019006
39. Banister D (2019) Beyond mobility: planning cities for people and places. Transp Rev 39(3):418–419

Chapter 3
Youth-Led Response to COVID-19 in Informal Settlements: Case Study of Mathare, Nairobi, Kenya

Douglas Ragan, Olga Tsaplina, Amandine Mure-Ravaud, Mary Hiuhu, Isaac Muasa, and Kui Cai

3.1 Introduction

The destabilizing effect of COVID-19 pandemic is widely discussed in literature, public addresses and press, with the social and spatial dimensions of its spread and impacts almost immediately having emerged on surface. The cities were soon labelled as 'frontlines' of the fight against the deadly infection. For as many as 1 billion people residing in slums and informal settlements [4, 5, 29, 30], implementation of the public health guidelines issued by the World Health Organization (WHO) was close to impossible, given the lack of data, as well as aggravating access to water, sanitation, health, basic services, and housing. Vulnerable populations living in these

---

D. Ragan (✉) · O. Tsaplina · A. Mure-Ravaud · M. Hiuhu · I. Muasa · K. Cai
UN-Habitat, Nairobi, Kenya
e-mail: douglas.ragan@un.org

O. Tsaplina
e-mail: olga.tsaplina@un.org

M. Hiuhu
e-mail: mary.hiuhu@un.org

I. Muasa
e-mail: mwasaisaac@yahoo.com

K. Cai
e-mail: alyssa.cai@outlook.com

I. Muasa
Mathare Environmental (MECYG), Nairobi, Kenya

K. Cai
UN Volunteer, Beijing, China

© The Author(s), under exclusive license to Springer Nature Singapore Pte Ltd. 2023
L. Zhang et al. (eds.), *The City in an Era of Cascading Risks*, City Development: Issues and Best Practices, https://doi.org/10.1007/978-981-99-2050-1_3

areas are most at risk to contract the infection, suffer from the most severe cases of illness and socio-economic consequences of the pandemic.

In Kenya, the situation is made more critical by the fact that the country has a notable number of people living in informal settings. Since 2009, when 54.7% of Kenya's urban population was estimated to be living in informal settlements, amidst the slight decrease in the proportion of people living in slums, the absolute number continues to grow. Nairobi hosts some of the largest slums in the country, including Mathare, which is home to over 60,000 inhabitants per square kilometer. The population in these informal settlements is largely composed of young people, who generally constitute over 35% of the total Kenyan population for those between the age of 15–34 [33].

Young people make up a large proportion of the world's population and are among the hardest hit by this pandemic. Yet they are also an incredibly dynamic human resource, and thus—the backbone of any community strategy. The concept of youth-led development places young people at the center of their own and their communities' development, "moving youth from passive receptors of development to agents of positive change" [31]. Therefore, policy responses must include youth groups and recognize young people in decision-making to ensure positive change.

Confronted by the spread of the disease, like it was shown in the past experiences of fighting the epidemics, community-led initiatives have proven to be successful in mobilizing their internal resources to prevent the virus' spread [4, 5, 29, 30]. Originating complex systems operating in a highly uncertain environment, the effectiveness of the community-led interventions varies dramatically into the crisis. However, some similarities in the responses have proven to be successful and can be identified, creating a 'community of practice' amongst stakeholders spearheading the process.

This article aims to make an argument that young people, when fully and meaningfully engaged and put in leadership positions, contribute to building the resilience of their communities and drive a collective response to emergency situations like the pandemic.

UN-Habitat works in partnership with youth and youth organizations in informal settlements, particularly through the support provided to the One Stop Youth Centers and the Urban Youth Fund. This work is also part of UN-Habitat's COVID-19 Response Plan, which has allowed UN-Habitat to work hand in hand with the wider UN family, national and local governments, and communities, in providing immediate support in 17 countries.

This paper presents the case study of youth-led community response to the spread of COVID-19 in the informal settlements of Nairobi, Kenya, focusing on Mathare—the second biggest slum in Nairobi. The COVID-19 Youth-led Emergency Response project (March 2020–March 2021) that was carried out by local youth groups with the support of development actors can be used as a reflection material in responding to COVID-19 pandemic and other disasters globally. With many countries facing consecutive waves of the virus, there's a need to adhere to the WHO guidelines and facilitate the continuation of initiatives that empower vulnerable groups, especially in informal settlements.

The objectives of the paper are as follows:

- Analyze what key characteristics of youth and youth-led organizations influenced the effective response to COVID-19 pandemic in Mathare.
- Identify lessons learned from this youth-led community response, and
- Derive implications for bigger policy questions about public health responses.

From March 2020 onwards, UN-Habitat worked hand in hand with young people in informal settlements to provide an effective and multi-partner response to the pandemic. To be able to trace back the process of building up the response, several semi-structured interviews were conducted with the core people from the projects, while data gathered in time-series surveys in Mathare have been used as supporting evidence.

The main body of this paper is composed of four sections. First, through a literature review, we identify the effect of the pandemic on vulnerable communities, with the primary focus on informal settlements. The second chapter deals with the case study of Mathare's youth response to the public health crisis, its achievements, and challenges. The conclusion outlines the findings and potential areas of further research.

3.2 Community-Led Response to COVID-19 Crisis: Why Youth Matter

The sudden emergence of COVID-19 in early 2020 caught the world off guard; since approximately 90% of all cases were reported in urban areas (as of July 2020), cities and towns have become the center of attention for prevention and control measures. The effectiveness of these measures depends on the inherent ability of a system to respond and adjust, including imposing restrictions, tracking the spread of virus, conducting effective awareness raising campaigns and so on.

In urban areas where informality and inadequate living conditions exist, local communities and authorities do not display enough mutual trust, which would otherwise foster adherence to the new rules. And as the core learning gathered from the humanitarian response to the epidemic of Ebola in West Africa in 2014–2015, in settings where such trust is missing, where horizontal communications are not necessarily in place, it is of vital importance to properly engage with existing socially coherent local communities, i.e. reach out to community leaders that can lead the response, as well as support grass-roots self-mobilization efforts [14]. Provided that external actors such as international organizations and NGOs take on a supportive role of providing funds and requested training, the principle of ownership of a self-organized community is more likely to be upheld.

The inequalities and differences in contexts lay the groundwork for the disparity in responses. While new technologies are used, with the best ones known being ones that trace contacts to track the spread of the virus across the community to more experimental ones that can recognize the symptoms of being infected by COVID-19 by an owner's facial expression, the overwhelming majority of those living in informal settings do not possess these.

In contexts where strong preparedness mechanisms are hard to implement, reliable transmission and other data are not available, conducting communication and awareness raising activities are challenging, and the level of trust towards authorities is low, like in case of informal settlements, it was more likely to be observed that civil society and individual advocates take over the initiative. One example is in Liberia where youth have undertaken COVID-19 measures and initiated a grass-root initiative that has grown to be recognized as part and parcel of the nation-wide COVID-19 prevention mechanism [4, 5, 29, 30]. Some evidence from Ethiopia suggests increased effectiveness of youth mobilization for the success of the COVID-19 response [15]. International organizations such as UNHCR have shared inspiring examples of community-based responses to the pandemic, including youth-focused ones [34].

The pandemic disproportionately impacted different population groups, hitting hardest the underprivileged. Structural inequalities proved to be exacerbated by the pandemic. The multilayered intersectionality of destitution, precarious employment, inadequate housing and supply of basic services, youth, and seniors, and migrant or indefinite status populations such as slum dwellers, also impacted groups disproportionately. Due to precarious living conditions, slum dwellers are more prone to contracting infections because of malnutrition, lack of access to healthcare services and weakened immune systems [13].

The impact on the health of people residing in informal settlements and other precarious settings is more significant as they are at high risk both physically and mentally due to the worsening conditions. In terms of housing conditions, informal settlements are densely packed, oftentimes with several people sharing a poorly ventilated room, making it difficult to practice physical distance. Preventive sanitary measures, such as hand washing, are equally difficult, given the lack of running water and basic facilities. Finally, since a large majority of the workforce is informally employed, and most cannot work from home, thus requiring them to use overcrowded public transport which also makes social distancing impractical [4, 5, 29, 30].

At the same time, when the first data on COVID-19 impacts on different age groups emerged, governments of various levels, international organizations and non-governmental nonprofits have found that the lesser susceptibility of young people to the virus can be capitalized on to build up a stronger public response as a supplementary tool for either facilitating governmental response, or shifting the burden of it. Countries which undertook mobilization mechanisms have been quick to employ youth volunteers in areas such as reaching out to the elderly, who, being critically susceptible to the virus, have suffered most from the restriction of movement.

In contexts where strong preparedness mechanisms are hard to implement and reliable transmission and other data is not available, conducting communication and awareness raising activities are challenging, and the level of trust towards authorities is low, like in case of informal settlements, it was more likely to be observe how civil society and individual advocates take over the initiative. An example of the Liberian youth movement against COVID-19 that started as a grass-root initiative, grew to be recognized as part and parcel of the nation-wide COVID-19 prevention mechanism [4, 5, 29, 30]. Some evidence from Ethiopia suggests increased effectiveness of youth mobilization for the success of the COVID-19 response [15]. Some international organizations with great presence on the ground like UNHCR, share the lists of inspiring examples of community-based responses to the pandemic, including youth-focused ones [34].

Moreover, engaging youth in crisis response matters because it does matter to them. Although young people are less likely than older people to develop severe forms of COVID-19, they are an age group hit hard by the long-term socio-economic effects of the pandemic [22–26]. The negative impacts of the virus are exacerbated in contexts of fragility, emergency, disaster risk reduction and conflict, where social cohesion is already undermined, and institutional capacities and services are limited. In a global wellbeing survey conducted by UN-Habitat and Fondation Botnar between August and December 2020, as many as 70% of young people agreed or strongly agreed with the statement 'COVID-19 pandemic has impacted your everyday life routines', with 64% of all the respondents sharing that their life has changed for the worse. In these circumstances, being involved in community response to the pandemic, be it locally based production of protective equipment or sanitizers, income from servicing handwashing stations or creating awareness raising materials, was a way of making their ends meet for many young people.

The same survey pointed out that almost 14% of the surveyed young people connected their negative experience of the pandemic with feelings of isolation and resulting loneliness. It is likely that school shutdowns (27% of all the responses), as well as decreased opportunities to spend the spare time outside the house with their friends contributed to this feeling. A significant number of young respondents linked COVID-19 to positive influences in their life, sharing that the pandemic and associated restrictions helped them discover more opportunities, bond with their families and/or acquire the feeling of being connected to the rest of the world. Participating in or in some cases leading a community response to COVID-19 pandemic was found to be one of the ways to overcome this impression of being excluded.

On top of that, young people globally have disproportionately suffered from the effects the pandemic has had on education. Due to fear of community spread of the COVID-19 virus, educational institutions were closed during the epidemic, depriving children, and youth of their education. Despite the adoption of e-learning and the use of online platforms to continue the curriculum, it was difficult or often impossible for young people in informal settlements to access online classes due to lack of internet access or Wi-Fi access points, electricity, and infrastructure.

The economic and social effects of the health crisis, as well as travel restrictions and social isolation, have led to a significant increase in domestic violence. An increase in violence against young women and girls has been reported worldwide, leading to the exacerbation of the 'Shadow Pandemic' as observed by Phumzile Mlambo-Ngcuka, UN Women Executive Director. In informal settlements, many young people are forced to stay for extended periods of time in their homes with their abusers. Due to the limitation of services during the pandemic, support services for victims have sometimes been interrupted or inaccessible.

In times of nation-wide imposed social distancing rules and curfews, youth have also been increasingly targeted by police and other law enforcement officials [1]. In a context of increasing police violence to regulate curfews and containment directives, young people in informal settlements are particularly vulnerable. As the conditions in which they live hinder the application of curfew and confinement measures, they are often more targeted by police control operations. In Nairobi, violent and coercive enforcement of emergency measures is omnipresent, particularly in informal settlements. The United Nations Human Rights Office of the High Commissioner's monitoring work with the Mathare Social Justice Centre have documented widespread police violence, including beatings, the use of live ammunition and tear gas, sexual violence, and damage to property [2].

UN-Habitat found that young people in Mathare reported mental health in the community, education, gender-based violence, and police brutality were named as first priorities for any planned intervention. Once given an opportunity to voice their concerns and do something to address them, youth faced the challenges headon. Providing evidence on how creative young people can be in responding to a crisis during the pandemic is what this paper is about. Some of the instances are documented through surveys run amongst young people globally since the beginning of the pandemic. Through the analysis, the youth response came down to them being mobile and able to reach out to the most vulnerable, their ability to launch social services and enterprises, and their utilization of technological solutions as an alternative to the traditional mechanisms of help.

During the pandemic, young people have been at the forefront of the fight against COVID-19 by being innovative and flexible in their response. For example, some young people have turned to the production and sale of protective masks. Many young people are volunteering to fight the pandemic by raising awareness in their communities.

3.3 Case Description: Youth in Mathare Responding to COVID-19 Crisis

Mathare is one of the largest informal settlements in Nairobi, home to approximately 500,000–800,000 people [33]. This settlement hosts the Mathare Environmental One-Stop Youth Centre, established in Mlango Kubwa, a ward of the Mathare Informal Settlement. The One-Stop Youth Centre is supported by UN-Habitat to provide a space for information, training, and a platform for setting up projects by and with the youth of Mathare.

The Youth-led COVID-19 response was initiated by youth representatives of local grass-root organizations in early March 2020, through the leadership of Mathare Environmental Conservation Youth Group (MECYG), an association of local youth that started as a community solid waste collection body, which runs the One-Stop Youth Centre. "Leaders [among youth] are always concerned about their community", shared Isaac Mwasa, coordinator of MECYG, "It is so congested here—a breeding ground for the virus. We felt that a small act of kindness is needed, so we started looking for funds to establish handwashing stations around the community."

Youth representatives of MECYG and other youth-led associations in Mathare and other informal settlements across Nairobi faced similar challenges in regards to the pandemic. Recalling that at the very onset, young people were faced by the challenge of having to deal with fundraising and identifying and approaching donors. "We knew whom to reach out to but felt that we need some help in writing a proposal." shared Isaac Mwasa, who then adds that having access to partner organizations helped them enhance the capacities of his team.

A proposal was developed which included several strategies identified as critical for the success of the response, including tackling health impacts through sustaining hygiene, implementing a feeding programme, carrying out public awareness campaigns, addressing livelihood impacts through promoting project-serving employment of community members. The Proposal was written initially for Mathare, and was then expanded to incorporate youth groups in other counties in Kenya, and countries such as Somalia and Ecuador.

Sanitation and hygiene were prioritized in the early stage of the response. MECYG submitted the proposal to and received support from the Canadian High Commission and the Norwegian embassy in Nairobi and private sector donors such as the Victor Wanyama and Chandaria Foundations. Youth from the community were consulted on substantive matters on the design and implementation of the programmes, while communications with donors were handled by the UN-Habitat team working in youth programming.

Initially, as the support from donors was under review, two UN-Habitat-sponsored handwashing pilot sites were set up in late March and achieved 8,000 handwashes in 10 days. Based on the pilot experience, UN-Habitat and MECYG designed capacity-building courses, comprised of training on handwashing knowledge and self-protection for site assistants, and made up of youth recruited from local youth-led organizations. Following the start of donor funding, twenty-six handwashing sites were established in Mathare. Each site was open from 6.00 am to 10.00 pm with a paid assistant organizing people into properly washing their hands and other sanitary-hygiene best practices.

The next step was to scale up the scope of activities, for which MECYG reached out to other youth-led groups, totaling 33 over the next few months, each having their own geographical and subject area of responsibility. The selection of youth organizations was made by MECYG based on previous experience of collaboration, including groups of local artists for awareness raising campaigns, environmental and community development groups (e.g. Kambi Saafi, Mathare Roots), women's groups (e.g., Wonderful Mothers), professional associations (e.g., Mlango Kubwa Boda-Boda Group, which is composed of motorcycle drivers), and groups from other communities such as the Kibera informal settlement, Eastleigh and Nyeri and Mandera Counties.

The interaction with these grassroot groups as entry points to access the community was instrumental for the success of the response. The unity of goals led to the unity of communications amongst the groups, as streamlined by MECYG and leaders of the participating organizations, forming an 'umbrella ad-hoc committee of emergency response'—which have served as basis for a network gradually growing into the Youth-led COVID-19 Emergency Response Coalition—young people pulled together their human resources to compensate where complimentary funds, space and physical assets could be missing.

Co-led by MECYG and Kibera Community Emergency Response Team, by the middle of the year 2020, the Coalition established over 90 handwashing stations in strategic high traffic places in their respective communities. UN-Habitat performed an advisory role to these groups, while Embassies and the private sector provided financial support. The local government provided logistical support on the ground.

Regarding the management of the health crisis, apart from the pressing need of access to clean water and sanitizers, the socio-economic crisis provoked by the pandemic and associated restrictions made residents struggle with acquiring masks, which were now required to be worn in public places. The following proposed step came as a response to the changes in the national legislation meant to curb the spread of the disease. In Kenya, a person with no mask could be denied a entry into a shopping mall or service delivery area or could be arrested and/or fined. A solution was found through donations from charitable groups such as the Victor Wanyama Foundation, led by football player and UN-Habitat Goodwill Ambassador Victor Wanyama, who donated funds needed to cover the costs associated with production of 10,000 masks for Mathare informal settlement sewn by local female tailors. As Gerry Wonderkid, representative of Mathare Empire put it, 'the impact was visible. I attended launches of Victor Wanyama initiative in Mathare and Kibera and saw a great turnout and felt how the people seemed empowered'. The local youth distributed reusable masks which were much cheaper than disposable surgical masks.

To further address the impacts on livelihoods, the response of the youth of Mathare and its partners poised to promote employment and to create job opportunities in COVID-19 response projects for running the handwashing sites, make fabric masks and support the food delivery. For the maintenance of the handwashing stations, the assistants were paid a daily allowance and provided with personal protective equipment kits. The post-COVID-19 recovery challenge was to reintegrate youth back into the disrupted labour market, making the achievements of the programme more durable.

Another strong component of the response was education campaigns through workshops, leafleting, and creating murals. People in informal settlements have limited access to reliable information on health care, such as official information from television. With limited access to education and therefore low levels of literacy in the context of disease prevention, it is more difficult for them to distinguish false information from correct knowledge about infection control. Youth responded by creating a YouTube video blog series. "Mathare Dispatch" became a series of six videos in English and Swahili on COVID-19 recorded by local youth videographers that educated the community on COVID-19 prevention and combatted misconceptions and misinformation. These videos were then shown at MECYG, community centres and disseminated online to a larger audience.

Understanding that the spread of accurate information through internet-based media might be limited, Mathare youth mobilized community radio to reach out to those who do not have regular access to broadband. Later, as the handwashing stations assistants conducted the survey among the members of the community, they found out that nearly 70% of all respondents got the information about COVID-19 from the local radio broadcasts.

In addition, one of the local coordinating groups, Mathare Roots, brought together local young artists to paint graffiti on the walls of the main entrances to the local community to disseminate COVID-19 prevention information, public health and messages on other issues related to the pandemic such as police brutality and gender-based violence. Many of the COVID-19 prevention mural gained international renown, with articles being written in the Guardian Newspaper https://www.theguardian.com/world/video/2020/may/04/the-coronavirus-murals-trying-to-keep-kenyas-slums-safe-video and the BBC https://www.bbc.com/news/world-africa-53150397.

In the medium term, an extensive community mapping was carried out to map out the existing healthcare facilities and main entrance of communities, for informing future prevention and response measures. https://youtu.be/0iSXMFfkimM. This mapping work was done by the young people, in partnership with UN-Habitat's Research Unit, who provided technical assistance, enabling a tailored response to the existing situation. This consists of geospatial mapping to identify existing assets such as health care facilities, clinics, and water stations.

In December 2020, the handwashing station assistants were trained by UN-Habitat to conduct a survey among the members of the community, to map out the degree of vulnerability of selected groups and generally understand the impacts of the pandemic on the community. The questionnaire was co-designed by local youth and UN-Habitat to include questions that were deemed most relevant to the realities of Mathare amid the pandemic. The survey also allowed to receive some preliminary results of the intervention.

3.4 Case Analysis

The required rapidity to dispatch measures to curb the growing public health crisis, which, if unaddressed urgently, endanger lives of thousands of residents of Mathare, left no alternative but to rely on the community to respond. Considering this, we argue that it is the leadership of local youth as exemplified by MECYG that made the response a reality. Founded in 1997 MECYG facilitated the removal of barriers to information dissemination associated with the distrust towards authorities and 'outsiders'. This trust was also supported by the fact that local youth were actively engaged in the project at all its stages, from the design, through the implementation and monitoring stages all the way to evaluation of its outcomes. Despite the difficulties related to the lack of resources, the young people adopted and implemented specific intervention strategies to respond to the most urgent needs, as they saw them in the community, that focus mainly on the health impacts, the livelihood impacts and impacts on resources and service ability.

There are several characteristics of youth that enabled their agency that eventually defined the effectiveness of the response. The dynamism of Mathare youth, their ability to empathize, connect and network beyond community borders allowed the response to more effective scale-up, leading to the formation of the Youth-led COVID-19 Emergency Response Coalition that helped increase the number of communities covered by the intervention. That is why youth mobilization focused on communication and awareness-raising activities within the community was critical in awareness raising. Also, being the most dynamic demographic, early adopters, and digital natives, appearing at the edge between their home community and the international world, as represented by partner international organizations (in this case UN-Habitat), youth played an essential part in norm internalization across their communities. The instrumentality of community-led response headed by youth is conditioned by the fact that youth, seen as part of a community and not outsiders, step in as entrepreneurs and set a pattern for the rest of the community to follow (e.g., in case of setting handwashing stations and spreading the awareness via murals).

Another point that enabled the involved parties to effectively dispatch the response was the knowledge of the physical community space that local young people possessed and deepened in the course of community mapping exercise. The handwashing stations, placed in strategic locations in the neighborhood such as the main entry points, provided a public amenity for the population of Mlango Kubwa. These public amenities are managed by young people, creating job opportunities for them, and providing the population with information about the management of the health crisis. And while many healthcare facilities can be easily recognized and pointed out by locals, they often did not have the capacity to perform public health functions such as handwashing. The need for broader public health infrastructure to fend off COVID-19 was clearly highlighted—for example, in case of adequate sanitation facilities, which are often missing in informal settlements.

Economic crisis associated with the pandemic created a demand for new opportunities to help ends meet among locals, who were eager to partake in implementing the response measures to support their livelihoods. This points out the necessity to ensure that the primary motivations of the beneficiaries and youth directly involved in implementation are adequately addressed in a response project's budget.

When it comes to informal settlements, as shown by the example of the Mathare case, grassroots organizations often have limited access to external resources. In this respect, an effectively equitable multi-stakeholder partnership between international organizations, non-governmental bodies and development actors, the private sector, local and national government, and youth and grassroots organizations are an essential prerequisite of a successful community-led response. The existence of a One-Stop center as a structure run by MECYG was beneficial for capturing resources and acts as a platform for incubating youth-led development projects. This creates more opportunities and solutions for young people in informal settlements to enable a transformative change in their communities.

As the experience drawn interventions shows, many challenges that can be encountered in this context are of a structural nature. For instance, the necessary infrastructure to rapidly undertake an effective public health response is often missing, including healthcare, water and sanitation, food supply, education, violence prevention and so on. Secondly, communication channels between an informal settlement and the rest of the world are disrupted causing the atmosphere of fear and distrust due to the long-standing record of unsatisfactory relationships between locals and 'outsiders'. Therefore, it is practically impossible to achieve an effective outreach without the young people who play a role of 'entry points' for any intervention, which makes them powerful agents of change in their communities. The engagement of local and national governments is also critical, as they provide critical support through infrastructure, training and regulatory permissions. Other difficulties in implementing the COVID-19 response program alongside young people from the community include tradeoff between community service and income generation they must make and lack of reliable data about the spread of virus and its consequences.

While response measures are being gradually transitioned towards post-COVID-19 recovery, youth must be among both beneficiaries and agencies of this recovery process, an engagement of young women and men in the recovery as co-creators. This is more crucial in informal settlements as youth-led and community-driven organizations have often taken over the provision of services considered to be the domain of local authorities, such as waste collection and recycling or the management of public spaces [33].

Furthermore, the skills relevant to emergency response that young people receive are not rendered useless in times when the crisis seems to be averted. They become the backbone of an emergency prevention response that can be enacted in times of other crises such as future pandemics or natural disasters. Moreover, through triggering a set of actions for resource mobilization and partnership building, Mathare youth were able to engage with key agencies, whether governmental (the Government of Kenya), international (UN-Habitat) or private sector (Victor Wanyama Foundation), forming strategic partnerships for the success of future interventions, to the point of forming a youth-led COVID-19 global coalition. The long-lasting impact of this movement should be reflected upon in future research.

The example of youth in informal settlements shows us how important a youth-led response to the pandemic is in building community resilience. The question remains, to what extent we can help sustain this response over time. Indeed, if the long-term effects of the crisis are to be tackled, youth-led responses need to be both sustainable and durable.

3.5 Conclusions and Lessons Learned

Approximately one third of the global population are living in informal settlements, where high levels of congestion, lack of adequate sanitation, precarious housing makes the spread of virus especially hard to control, leading to recurrent waves, and threatening to transform informal settlements into dangerous points of virus transmission. As disease experts warn, slum dwellers are more prone to contracting infections because of malnutrition, lack of access to healthcare services and weakened immune systems. Furthermore, the spread of misinformation is equally dangerous as it spreads due to lack of trust towards authorities and missing access to correct information about the virus. The protocols are also not likely to be observed if adherence to them would endanger slum dwellers' livelihoods. Lastly, lack of funds necessary for survival, increased mental pressure, movement restrictions, including those forcing women, children, and other vulnerable groups of population to stay within four walls, account for the peaked violence, unchecked in informal conditions.

As it can be seen from the analysis of the case of Mathare, the argument that an emergency response is effective when the community is onboard is well evidenced. The most important factor of success is the availability of essential and irreplaceable assets that youth associations had on the ground. First, having a more than 20 years history of engaging with the members of community, other grassroot organizations and international partners, and yet remaining a youth-led entity, MECYG has become a critical entry point for the entire emergency response project that otherwise would lack in trust of the community members, generally suspicious of outsiders. Secondly, this credibility holds as MECYG and broader community youth have been meaningfully engaged in project design, its implementation, and evaluation of the outcomes. Thirdly, development actors and international organizations were playing a role of enablers by providing necessary funds and capacity building activities to these organizations. And finally, the actual economic needs of the youth involved in the implementation of the project should be taken into account to ensure more long-lasting success.

The success of young people's involvement in the local response to COVID-19 is further evidence of the need for them to have the opportunities to engage with local, regional, national and international agencies focused on public health and wider policies of community resilience building. Young people play a frontline role as drivers of community resilience and social inclusion. In the informal settlements in Nairobi, the youth demonstrated a strong capacity to organize and mobilize for the immediate management of the pandemic, which once again illustrates the responsiveness and effectiveness of youth organizations in crisis management. Local and national governments should identify and encourage such bottom-up youth-led actions, rather than responding to crisis situations with top-down strategies.

Youth citizenship and engagement in governance goes beyond political engagement but also involves the ability to secure access to economic resources, basic services, and socio-political assets. This begins with engaging young people as equal partners, acknowledging their rights to participate in decision-making processes

[32], as *'Building back better'* would require seizing the momentum of community engagement to mobilize public action towards implementing SDGs [22–26].

The key lesson learned is that community-led responses planned, initiated and executed by grass-root associations, particularly youth groups, as exemplified by the youth-led emergency project in Mathare, remain useful in responding to COVID-19 and similar pandemics in circumstances where institutions are weak, and informality persists. Complementing this finding by the evidence from the past outbreaks in highly congested urban areas, supporting community-based initiatives, not leading, or patronizing them, is more likely to ensure the success of prevention measures.

References

Sources of Materials and Online Publication

1. Human Rights Watch (2020) Kenya: police brutality during curfew. https://www.hrw.org/news/2020/04/22/kenya-police-brutality-during-curfew. Accessed 6 June 2021
2. Office of the United Nations High Commissioner for Human Rights (OHCHR) (2020) Kenya: monitoring human rights impacts of COVID-19 in information settlements (15 April–6 May 2020). https://www.ohchr.org/Documents/Countries/KE/15april-6may-OHCHR-SJCWG.pdf. Accessed 14 June 2020
3. Srivastava R (2021) As COVID-19 ravages India, a slum succeeds in turning the tide. https://news.trust.org/item/20210511225252-652st/?fbclid=IwAR0XKUcvQ2wEtmC9Au1X0To769wP-fmJfjbhP9ytrTDQ_PZo48xa7Pqtxhw. Accessed 6 June 2021
4. UN-Habitat (2020) Fighting COVID-19 in Africa's informal settlements. https://unhabitat.org/fighting-COVID-19-in-africa%E2%80%99s-informal-settlements. Accessed 14 June 2021
5. UN-Habitat (2020) Youth in informal settlements take the lead on fight against the spread of COVID-19. https://unhabitat.org/youth-in-informal-settlements-take-the-lead-on-fight-against-the-spread-of-COVID-19. Accessed 14 June 2021
6. UN-Habitat (2021) Kenya habitat country programme document (2018–2021). https://unhabitat.org/sites/default/files/2019/09/hcpd_kenya_2018_-_2021_0.pdf. Accessed 6 June 2021
7. UN-Habitat Youth (2021) COVID-19 and youth in informal settlements: engagement for global action. In: Report on the virtual summit on the occasion of the international youth day 2020. https://www.unhabitatyouth.org/COVID-19-and-youth-in-informal-settlements/. Accessed 6 June 2021
8. UN Women (2020) Violence against women and girls: the shadow pandemic. Statement by Phumzile Mlambo-Ngcuka, Executive Director of UN Women. https://www.unwomen.org/en/news/stories/2020/4/statement-ed-phumzile-violence-against-women-during-pandemic. Accessed 10 June 2021
9. Wickramanayake J (2020) Meet 10 young people leading the COVID-19 response in their communities. https://jayathmadw.medium.com/meet-10-young-people-leading-the-COVID-19-response-in-their-communities-685a0829bba8. Accessed 22 April 2021
10. Wickramanayake J (2020) Meet 10 young people leading the COVID-19 response in their communities, 2nd edn. https://jayathmadw.medium.com/10-young-people-fighting-COVID-2nd-edition-218a741e611b. Accessed 22 April 2021
11. Women Deliver (2020) 10 ways young people are leading the way against COVID-19. https://womendeliver.org/2020/10-ways-young-people-are-leading-the-way-against-COVID-19/. Accessed 22 April 2021

Literature

12. Ballard PJ, Hoyt LT, Pachucki MC (2019) Impacts of adolescent and young adult civic engagement on health and socioeconomic status in adulthood. Child Dev 90:1138–1154
13. Bourke C, Berkley J, Prendergast A (2016) Immune dysfunction as a cause and consequence of malnutrition. Trends Immunol 37(6):386–398. https://www.ncbi.nlm.nih.gov/pmc/articles/PMC4889773/. Accessed 10 June 2020
14. Campbell L, Miranda Morel L (2017) Learning from the Ebola response in cities: communication and engagement (ALNAP Working Paper). ALNAP/ODI, London. https://www.alnap.org/help-library/learning-from-the-ebola-response-in-cities-communication-and-engagement. Accessed 5 June 2021
15. Getaneh Y. Yizengaw A, Adane S, et al (2020) Global lessons and potential strategies in combating COVID-19 pandemic in Ethiopia: systematic review. https://doi.org/10.1101/2020.05.23.20111062v1. Accessed 30 Nov 2021
16. Harrouk C (2021) 12 Key principles for an effective urban response during COVID-19. https://www.archdaily.com/961166/12-key-principles-for-an-effective-urban-response-during-COVID-19. Accessed 7 May 2021
17. Holmes EA, O'Connor RC, Perry VH, Tracey I, Wessely S, Arseneault L, Bullmore E (2020) Multidisciplinary research priorities for the COVID-19 pandemic: a call for action for mental health science. Lancet Psychiatry 7:547–560. https://doi.org/10.1016/S2215-0366(20)30168-1
18. OECD (2020) Youth and COVID-19: response, recovery and resilience. https://read.oecd-ilibrary.org/view/?ref=134_134356-ud5kox3g26&title=Youth-and-COVID-19-Response-Recovery-and-Resilience&_ga=2.261025677.1446400989.1622919592-489222429.1622305172. Accessed 6 June 2021
19. Pavarini G, Lyreskog D, Manku K, Musesengwa R, Singh I (2020) Debate: promoting capabilities for young people's agency in the COVID-19 outbreak. Child Adolesc Mental Health. https://doi.org/10.1111/camh.12409
20. Sen A (1993) Capability and well-being. In: Nussbaum M, Sen A (eds) The quality of life. Clarendon Press, Oxford, UK, pp 30–53
21. United Nations (2020) Shared responsibility, global solidarity: responding to the socioeconomic impacts of COVID-19: report of the secretary-general. https://unsdg.un.org/sites/default/files/2020-03/SG-Report-Socio-Economic-Impact-of-COVID19.pdf. Accessed 6 June 2021
22. United Nations (2020) A UN framework for the immediate socio-economic response to COVID-19. https://www.un.org/sites/un2.un.org/files/un_framework_report_on_COVID-19.pdf. Accessed 7 May 2021
23. United Nations (2020) COVID-19 and human rights. We are all in this together. https://www.un.org/sites/un2.un.org/files/un_policy_brief_on_human_rights_and_COVID_23_april_2020.pdf. Accessed 7 May 2021
24. United Nations (2020) COVID-19 in an urban world. Policy Brief. https://www.un.org/sites/un2.un.org/files/sg_policy_brief_COVID_urban_world_july_2020.pdf. Accessed 10 May 2022
25. United Nations (2020) The United Nations comprehensive response to COVID-19: saving lives, protecting societies, recovering better. https://www.un.org/sites/un2.un.org/files/un_comprehensive_response_to_COVID-19_june_2020.pdf. Accessed 6 June 2021
26. United Nations Inter-Agency Network on Youth Development (2020) Statement on COVID-19 and youth. https://www.un.org/development/desa/youth/wp-content/uploads/sites/21/2020/04/IAYND-Statement-COVID19-Youth.pdf. Accessed 14 June 2021
27. UN-Habitat (2016) Slum almanac 2015–2016: tackling improvement in the lives of slum dwellers. https://unhabitat.org/sites/default/files/documents/2019-05/slum_almanac_2015-2016_psup.pdf. Accessed 14 June 2021
28. UN-Habitat (2020) UN-Habitat COVID-19 response plan, Nairobi, Kenya. https://unhabitat.org/sites/default/files/2020/04/final_un-habitat_COVID-19_response_plan.pdf. Accessed 6 June 2021

29. UN-Habitat (2020) World cities report: the value of sustainable urbanisation. https://unhabitat.org/sites/default/files/2020/10/wcr_2020_report.pdf. Accessed 10 June 2021
30. UN-Habitat (2021) Cities and pandemics: towards a more just, green and healthy future. https://unhabitat.org/sites/default/files/2021/03/cities_and_pandemics-towards_a_more_just_green_and_healthy_future_un-habitat_2021.pdf. Accessed 14 June 2021
31. UN-Habitat Youth (2012) Global youth-led development report series, Report 1 challenge and the promise of youth-led development. UN Habitat. https://www.unhabitatyouth.org/wp-content/uploads/2020/12/the-challenge-and-promise-of-youth-led-development-1.pdf. Accessed 14 June 2021
32. UN-Habitat (2013) Advancing youth civic engagement and human rights with young women and young men
33. UN-Habitat Youth and Tone Vesterhus (2015). Youth led development: a case study from the Mathare slum. https://www.unhabitatyouth.org/wp-content/uploads/2020/12/Youth-Led-Development_Mathare-Slum.pdf. Accessed 14 June 2021
34. UNHCR (2020) UNHCR and partner practices of community-based protection across sectors in the East and Horn of Africa and the Great Lakes Region. https://knowledgecommons.popcouncil.org/cgi/viewcontent.cgi?article=2311&context=departments_sbsr-rh. Accessed 30 November 2021

Chapter 4
COVID-19 in Angola's Slums: Providing Evidence and a Roadmap for Participatory Slum Upgrading in Pandemic Times

Miriam Ngombe, João Domingos, and Allan Cain

4.1 Introduction

4.1.1 The Pandemic and the Vulnerability of Slums in the South

90% of COVID-19 cases have been registered in cities, making the urban areas the disease's epicenter world-wide [32]. But within cities, the new coronavirus discriminates and not everyone is affected the same way. North to South, East to West, the pandemic has hit the poor and the marginalized the hardest [2]. Moreover, new research shows that since the global onset of the pandemic, COVID-19 has become a developing country pandemic [11], with cities in the Global South being particularly hit. Thanks to reporting and computer modeling, scientists now estimate that accounting for the unreported infection and death toll in the Global South this part of the world bears the brunt of not only the economic and social repercussions, but that it is worse affected than the Global North health-wise as well. The reasons for this are varied and already signaled during other infectious outbreaks. From the public health perspective, the lack of basic healthcare (including testing and vaccination), water and sanitation challenges, and overcrowding have historically created prime environments for the spread of infectious diseases, of which the new coronavirus is only the latest. Socio-economically, poverty and the absence of social protection drives millions to seek daily wages despite the health risks, making self-isolation, curfews,

---

M. Ngombe (✉) · J. Domingos · A. Cain
Development Workshop Angola, Luanda, Angola
e-mail: miriam.ngombe@gmail.com

A. Cain
e-mail: allan.devworks@angonet.org

and other restrictions difficult to comply with. The impact of other factors such as trust in government or belief in misinformation [21] need to be further studied.[1]

In this context, informal settlements in the Global South are the most at-risk of all zones, harboring the highest density of COVID-19 cases [25]. UN-Habitat identifies informal settlements as residential areas meeting three main criteria: (1) residents have no guarantee of ownership in relation to the land or houses they inhabit, "with modalities ranging from occupation to informal rent", [26] (2) these areas often lack, or are isolated from, formal basic services and city infrastructure, and (3) their housing "may not comply with current planning and construction regulations, is often located in geographically and environmentally hazardous areas, and may not have a municipal permit" [30]. The slum as the most impoverished form of informal settlement has been defined as "a deprived urban area, often not recognized and addressed as an integral or equal part of a city, town or human settlement, and whose inhabitants have inadequate housing and inadequate access to basic urban services," which include clean water, sanitation, adequate living space and secure tenure (UN-Habitat, PSUP Fact Sheet). In practice, the terms slum and informal settlement are often used interchangeably when referring to excluded and/or deprived residential areas. By UN-Habitat estimates, in 2016 every eighth person on the planet lived in a slum. This number is likely higher half a decade later, meaning that a significant proportion of humanity lives in areas that are particularly vulnerable to the spread of infectious disease and where addressing a contagion is particularly complex.

However, slum communities cannot become the scapegoats of authorities, media, and society at large in the fight against COVID-19. More than overcrowding, it is bad governance that makes informal settlements more prone to being hotspots due to lack of basic public services and due to poverty resulting from the marginalization of these populations [5]. The logic of dedensification that has been proposed by cities in the wake of the pandemic is counterintuitive in the context of the new shift to environmentally conscious models of urbanism that advocate for more compact, integrated and less polluting cities [28]. The latter should be preferred to urban strategies that, in the better case, ignore the crucial role of the millions of informal residents in the city's functioning, and in the worst-case lead to forced evictions and destructions of informal settlements while bringing no viable alternatives.

4.1.2 Participatory Slum Upgrading as a Mitigation Strategy

Experience from the Global South has shown that informal settlements are an enduring component of cities, and will continue to be, at least in the medium run. Instead of wishing slums to suddenly disappear through short-sighted demolitions,

---

[1] Development Workshop Angola, founded in 1981, is Angola's oldest NGO. Present on the ground throughout most of Angola's independence, civil war and reconstruction years, DW has been involved in both humanitarian and development initiatives. Its current focus is on challenges related to Angola's rapid urbanization, including land tenure, and water and sanitation.

all relevant urban stakeholders (from the government to investors to the local communities) should engage in an open exchange and develop locally adapted solutions to upgrade, formalize and fully integrate these areas but also to equip communities to respond appropriately to health emergencies such as the COVID-19 pandemic [13]. In parallel, pro-poor policies must be adopted to correct the rising inequalities in urban development that bring about slum creation in the first place, and cause urban dilapidation in marginalized communities in the North as well [12]. In a similar vein, UN-Habitat and partners have developed the Participatory Slum Upgrading Programme (PSUP), a framework that aims to "change the mindset" around slums and present them as dynamic hubs crucial in the city's functioning. In practical terms, the program places slum-dwellers at the center of slum upgrading by harnessing community knowledge and encouraging participatory decision-making. Bringing together various urban stakeholders, including the relevant slum communities, this model hopes to stimulate local ownership and bring in sustainable solutions, aiming for a more integrated and humane approach to slum upgrading in particular, and for a more sustainable and inclusive urbanization in general. In many ways, PSUP borrows from the Social Urbanism model tested in various Latin American cities [17]. More recently, PSUP has been oriented around the call in Goal 11 of the 2030 Agenda for Sustainable Development to "make cities and human settlements inclusive, safe, resilient and sustainable".[2] It also aligns itself with the New Urban Agenda, coined at the Habitat III conference in 2016. In this context, PSUP aims to reduce by 50% the number of people living in slums across the world by 2030, progressively formalizing informal settlements and advocating for policies that will prevent their emergence in the future (UN-Habitat, PSUP Fact Sheet).

As a development paradigm, the participatory slum upgrading approach has been criticized for its potential to gentrify upgraded slum areas, for its incremental approach [6] and because it can provoke clashes between the will of the community and the preferences of the governing bodies [23]. Nevertheless, PSUP's stress on inclusive, shared governance and sustainable urbanism makes it a hopeful alternative to centralized, government-led urban planning on the one hand, and the uncontrolled growth of cities on the other, both of which have coexisted in Southern cities with mixed results. In this article, we argue that PSUP is a potentially scalable model for community-led development in Angola. Here, a "hybrid space" has existed at the abovementioned nexus of informality and centralized planning in cities [4]. While evictions of slum dwellers for the purposes of property development are commonplace, slums have been allowed to swell and multiply in the absence of a coherent national policy as to low-income housing. Additionally, PSUP can be valuable in mitigating the current pandemic as it can be adapted to meeting public health needs in slums through direct involvement of their inhabitants (in awareness campaigns, emergency planning committees, water, sanitation and waste management teams, etc.) as we will explore later.

---

[2] https://www.globalgoals.org/11-sustainable-cities-and-communities, accessed on June 3, 2021.

4.1.3 The Angolan Musseques in Times of COVID-19

In Angola, the slum population estimates vary due to a lack of official statistics. According to the World Bank, about half of Angola's urban population lives in slums, down from 86% in 2005 when the country started recovering from decades of civil war.[3] Angola's 2016 National Report to Habitat III[4] put the figure at 63%, and Barros and Balsas [1] put the number of slum-dwellers in urban areas at 70%. Whatever the most accurate statistic, it is certain that slums have been a long-standing feature in most of Angola's urban centers, and their growth has symbolized the principal characteristic of Angola's urbanization process.

The situation remains particularly accentuated in the capital city Luanda, where inequalities are also the most poignant. In the past two decades since the end of the war, rich gated communities have sprung up next to sprawling slum neighborhoods, or sometimes on top of them (Table 4.1 and Fig. 4.1). Slums across Angolan cities have continued to mushroom and since the colonial era as a result of decades of unassisted or chaotic city development, the arrival of millions of internal refugees in the war years (1970s–2002), and due to the millions of mostly poor migrants arriving in search of opportunities in the post-war period. Additionally, high fertility rates of over 5.5 children per woman further drive the population increase in these settlements [15]. These high-density, informal areas have largely suffered from neglect when it comes to public policy and basic infrastructure projects. It is important to note, however, that access to quality urban housing in Angola is a nation-wide problem. According to the results from the 2014 General Census of Population and Housing, only 57% of Angolan households in urban areas had safe access to drinking water, only half (50.9%) had access to electricity, and just over a third (37.5%) had an appropriate solid waste management system. Nearly 90% of all Angolan settlements were self-built, a trend that poses a particular challenge to the national administration in meeting basic sanitation and infrastructure needs of all its citizens. With respect to slum settlements, various government-commissioned programs were created in recent years to streamline slum upgrading [31], yet tangible improvements are few and the musseques continue to remain on the periphery of public interest. More accent continue to be put on building large-scale, government-planned housing estates such as the famous Kilamba Kiaxi outside of Luanda, which are out of reach for most ordinary Angolans [3].

It is therefore from a place of accrued vulnerability that slum dwellers have had to face the outbreak of the new coronavirus as it hit Angola in March 2020. More than a year into the pandemic, the extent of the pandemic in the slums is little known when it comes to government-generated estimates. As of June 1st, 2021, the number of reported COVID-19 cases nation-wide stood at nearly 35 000, 792 deaths, and

---

[3] https://data.worldbank.org/indicator/EN.POP.SLUM.UR.ZS?locations=AO, accessed on June 11, 2021.

[4] https://www.angonet.org/dw/sites/default/files/online_lib_files/201407_dw_habitat_iii_report_for_ministry_of_urbanism_draft.pdf.

Table 4.1 Luanda settlement characteristics in numbers (2021)

Type of settlement		% of total population	2021 population
Informal settlements	Rural settlement	3	291,473
	Working-class neighborhood	1	65,733
	Social housing	3	259,966
	Old musseques	25	2,130,755
	Orderly musseques	9	794,354
	Peripheral musseques	28	2,443,975
	Transitional musseques	2	185,707
	Other	0.03	2,842
Formal settlements	Apartment buildings	23	1,995,212
	Colonial urban centre	1	104,654
	New suburbs	1	128,253
	Planned self-built construction	2	214,373
	Industrial zone	0.16	14,580
Total		100	8,631,876

28 000 recoveries, according to Angola's Interministerial Commission.[5] However, there has been much skepticism around these numbers as Angola is featured among the countries with the lowest rates of testing per capita.[6] At the time of writing in June 2021, testing is still not widely available across the country, and where it is, its prices are generally inaccessible for the ordinary citizen.[7] Moreover, the public release of data is often incomplete and when more specific information is available, it is often outdated. The only public database of COVID-19 cases shows the overall number of active infections, recoveries and deaths, but no information is available when it comes to specific cities or municipalities, age groups or gender break-down. For the common citizen, daily press releases are the only way of tracking which provinces registered new infections (generally only the eldest and the youngest infected patient get reported). As of June 1st, 2021, the latest breakdown of COVID-19 cases by

---

[5] https://www.cisp.gov.ao:10443/en/provincias/, accessed on June 1st, 2021.

[6] https://www.worldometers.info/coronavirus/, accessed on May 29, 2021.

[7] In June 2021, the cheapest individual test cost 6000 kwanza (cca $9.35), a prohibitive price for most of the population. Free tests are occasionally available when mass testing is done (in markets, place of work etc.), or when a person was in direct contact with a confirmed COVID-19 case, but often the test results would not be available even after 72 h.

province on the official government website dated back to January 2021. Most cases were registered in Luanda and in large urban centers. But consistent reporting on infections by municipality or neighborhood is rare, with only selected neighborhoods such as Hoji-ya-Henda in Luanda getting particular attention when they are placed in special sanitary perimeters.

In this generalized data scarcity, the slums are the most forgotten. Ironically, they might well be the most affected areas, not only by the infection rates, but also by the COVID-19 restrictions on the informal and formal economies, schooling, and adjacent issues such as price hikes of basic products. At various points in the pandemic, formal and informal trade was not allowed or only tolerated a few days

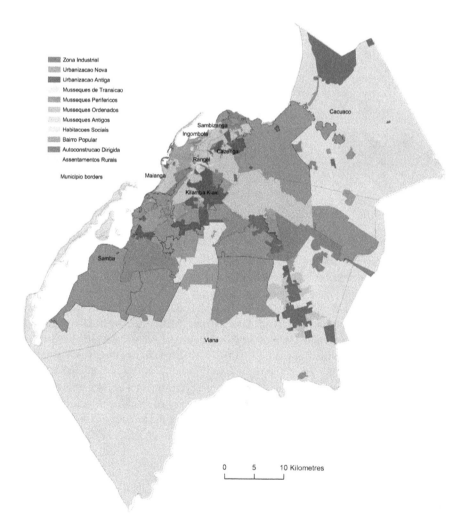

Fig. 4.1 Luanda's settlement characteristics—cartography (2021)

per week, slashing the incomes of many slum dwellers and causing difficulties in the search for basic goods and services for others. Public and informal transportation was either prohibited or their capacity and hours of circulation restricted, impacting the livelihoods of millions of slum dwellers working as taxi drivers or taking public transport to their place of work. Schools and daycare centers were shut in March 2020 and only started opening progressively between October 2020 and March 2021, impacting parents and school-going children for whom home-schooling is rarely an option. Child vaccination campaigns for polio and other infectious diseases were interrupted, and many pharmacies and healthcare facilities struggled to supply basic medicine and materials. The list of fallouts is long, but little precise data is available.

The scarcity of health and social statistics on slum dwellings and their populations has been identified as one of the greatest impediments to successful pandemic management worldwide [10], as well as a long-term obstacle to the drafting and implementation of appropriate policies related to informal settlements. There are exceptions to this trend in places such as Brazil, Kenya, South Africa and India where data-collection is abundant, even if not country-wide. The truth is, however, that a vast majority of crucial stakeholders have little to no reliable data on the state of the affairs in the world's slums. Researchers from the Sustainable Cities Survey highlighted that knowledge co-production is primordial to crafting successful slum improvement programs.[8] UN-Habitat has recognized the urgent need for up-to-date information gathering in the informal settlements and it has stressed its role as primordial within the Participatory Slum Upgrading Programme, ahead of designing any local projects.

One of the pillars of Development Workshop's (DW) work in Angola has been participatory data collection (on the community-level as well as through satellite imaging, remote sensing and GIS), not only to fill the void in reliable statistics on the Angolan urban areas, but also to conduct evidence-based advocacy and interventions. It is against this backdrop that DW led the Socio-Economic Survey on the Impact of COVID-19 on Slum Communities in Angola, adapting its existing data-collection programs across the country to survey the lived realities of slum dwellers during the pandemic. In parallel, DW has worked on implementing PSUP in slum communities with COVID-19 combat in mind, and more broadly with a mindset of sustainable, inclusive upgrading and governance. That is why the authors will draw out some of DW-PSUP practical initiatives and further recommendations, after highlighting a few main findings from the community study.

---

[8] https://www.datafirst.uct.ac.za/dataportal/index.php/catalog/832m, accessed on June 13, 2021.

4.2 The Socio-economic Survey on the Impact of COVID-19 on Slum Communities in Angola

4.2.1 Methodology

The DW-PSUP research study aimed to gather a holistic overview of the slum dwellers' well-being and living conditions during the pandemic, featuring multiple rubrics including housing structure and sanitation, household economics, health and healthcare access, and COVID-19 related questions. 1648 interviews were conducted telephonically between June-November 2021 across all of Angola's 18 provinces (Fig. 4.2). Based on the available public health data about the spread of COVID-19, the sample was weighed towards major cities. Nearly half (49%) of the interviewees were in the capital Luanda, and a third was from three other, populous provinces (Huambo, Benguela and Uíge). A small part of the participants was constituted of waterpoint caretakers who had been working with DW, while the rest were microcredit clients whose phone numbers were acquired from DW's sister organization, KixiCrédito. It is important here to note the study's bias towards the upper-income spectrum of the slum dwellers. First, in order to participate in the interview, the respondents had to own a phone, which is a resource that is often not readily available in the slums (due to the cost of the device and of airtime, but also due to the absence of electricity in many areas to recharge a phone). Secondly, many interviewees being microcredit clients, they had to dispose of a certain level of income. When public health restrictions are eased, additional in-person interviews should be performed to include a more representative sample of the population. Based on the reported location of the interviewees and on the low-income profiles, we assume that a significant majority live in informal settlements.[9] Women comprised 45% of all respondents. The margin of error based on the study sample was ±2.4%.

After a brief overview of general results (Table 4.2) we will highlight the findings that demonstrate the impact of COVID-19 and related public policy measures on the slum dwellers' material well-being. We will also propose a few recommendations and sketch out initiatives that DW has implemented as part of the Participatory Slum Upgrading Programme to demonstrate how various stakeholders in the urban environment can actively participate in finding sustainable, community-driven solutions in the slums during the COVID-19 pandemic and beyond.

---

[9] Because of official restrictions on movement due to the COVID-19 pandemic, we could not verify this hypothesis on the ground by interviewing within people's homes.

4 COVID-19 in Angola's Slums: Providing Evidence and a Roadmap …

Fig. 4.2 Study sample by province

Table 4.2 An overview of study results. *Source* DW project report to UN-Habitat, May 2020

1648 interviews [between June—November 2020; margin of error—±2.4%]	
Where?	
49%	In Luanda
27%	In three other provinces (Huambo, Benguela, Uíge)
Who?	
	45% : 55%
Housing	
	High density: 6.6 people and 2.6 rooms per household on average
	57% with reliable daily access to water
Employment and income	
	65% of men and women are self-employed, 21% have a regular job
	57% earn less than $80/month, 27% earn less than $40/month
	60% report a drop in income since pandemic started but impact on employment seems marginal
COVID-19 awareness	
	93% believe in the efficacy of COVID-19 prevention (masks, hygiene…)
	57% believe at least one fake news on COVID-19
	Main info sources—TV, radio, friends, Facebook
Health	
	1.3% had a COVID-19 case at home; 2.6% report a COVID-like death
	Malaria—most common sickness (1/5 of households); 2nd most common—high blood pressure

4.2.2 Key Findings

A Brief Overview

In sum, the study data revealed that the participating households house on average 6 people in 2 rooms, suggesting a certain level of crowding (Table 4.2). Only a third (36%) reported being capable of home-isolating a potential COVID-19 case in a separate space. Few households have reported a COVID-19 infection (1.3%), an observation that may be due to low testing, especially in the first months of the pandemic when the sample was gathered. One fifth (21%) reported at least one case of malaria at the time of interview. This rate would possibly have been higher had the study been conducted in the rainy months (December–April) when malaria cases tend to peak [22].

In a sample that is at 64% self-employed (we assume this means informal employment) and at 21% employed in a regular job,[10] employment has not dropped since the eve of the pandemic as drastically as some analysts have feared. On the other hand, the households' incomes have decreased drastically and expenses rose sharply, leaving a large proportion potentially food insecure and with reduced access to hygiene as they are left with less money to purchase soap, hand sanitizer and water. The government-instituted lockdown and restrictive measures seem to have weighed heavily on these households, of which a majority perform informal job activities and have reported virtually no official social protection to replace the loss of income. While many approve of mandates on masks or school-closing, most disagree with the closing down of markets and commerce.

Minimal Job Losses in Slums

According to the National Statistics Institute, employment at the end of 2020 had risen by 8.3% compared to a year ago [14]. This news was received with some skepticism,[11] seeing that COVID-19 destroyed jobs worldwide and Angolans continue to struggle to find jobs. However, this rise in employment could be explained by the soaring of informal labor, which accounted for 80% of all jobs in Angola, up from 72% in 2019 [15]. The report confirms a prediction by the World Bank that the Informal economy will have absorbed more workers, either due to jobs lost in the former sector (estimated at nearly half a million in 2020[12]), or as a result of those who have been trying to find formal jobs but gave up as the pandemic destroyed new opportunities in the formal sector [33]. Similarly, our study found that only 3% of participants lost employment since the start of the pandemic, likely due to the high levels of job informality of slum dwellers in general (Fig. 4.3). Based on our findings, 65% reported being "self-employed" in what we assume to be the informal economy. Neither category showed significant differences between genders. Further research would be needed in gauging precisely how many of respondents who labeled themselves as "employed" were employed in the informal economy (i.e., domestic workers, guards etc.) and how many self-employed people owned a formal license.

A third of respondents stated their specific employment (this question was added at a later stage in the study). Of these, over four fifths (76%) work in commerce, and an additional 12% are itinerant sellers. We will see later that this is likely why government restrictions placed on commercial activities (both formal and informal) were reported by the households as the most harmful.

---

[10] The following question about employment was asked: "What is your current working condition?" with the following options: (a) Employee (b) Self-employed (c) I lost my job during the pandemic (d) Unemployed already before the pandemic. No question was asked about the formality of the given employment.

[11] https://www.dw.com/pt-002/o-desemprego-est%C3%A1-mesmo-A-baixar-em-angola/a-565 33545, accessed on June 13, 2021.

[12] https://expansao.co.ao/artigo/144334/angola-perdeu-467-mil-postos-de-trabalho-formais-em-2020-?seccao=exp_mercm, accessed on June 13, 2021.

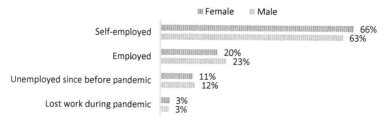

Fig. 4.3 Employment status of interviewees

Increased Hardship as Incomes Fall and Expenses Rise

Although employment among the study sample held constant since the onset of the pandemic (Fig. 4.3), households found themselves with lower income (Fig. 4.3). Our hypothesis leads us to believe that because of official restrictions on informal and formal commerce (e.g., limiting the number of days allowed for market or street sale) and on services (e.g., limited seating capacity in taxis and temporary bans on motorbike taxis), sales and hence incomes of these households decreased. The spillover effect is felt within the larger society in that with less disposable income families also spend less, taking the economy into a downward spiral.

It is interesting to note that the cities of Luanda and Benguela registered a more severe drop in incomes than the regional cities of Uíge (North) and Huambo (Center-South). One explanation for this can be that the enforcement of restrictions on the informal economy was greater in large metropolises, and especially in Luanda. Slum dwellers in the more rural provinces of Huambo and Uíge also have a greater access to alternative forms of subsistence such as small-scale urban or peri-urban farming, providing them with extra means of income when regular work opportunities are curtailed. The increase in urban agriculture initiatives during the pandemic deserves more academic attention. Among others, [27] reports how low-income families from Dhaka, Bangladesh turned to urban farming during the COVID-19 crisis to secure food.

Half of the study households have reported having spent more since the onset of the pandemic (Fig. 4.4). This can be the result of a combination of factors such as the hike in prices of basic goods (due to a severe drought which has made for poor harvests, because of rising inflation, and due to the devaluation of the national currency which makes particularly imported goods more expensive), the introduction of value added tax on goods and services a few months before the arrival of the pandemic, as well as due to restocking and transportation challenges during the pandemic. However, a third of participants reported having spent less, likely as a result of lower incomes but also as the nation-wide lockdown decreased spending on transportation etc. Similarly, studies from low-income communities in Bangladesh and in Brazil show a decrease in income and a rise in expenses, leading to accrued hardship for slum dwellers [18, 24].

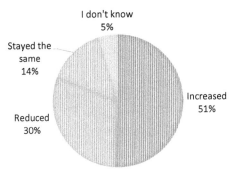

Fig. 4.4 Changes in household expenses since the beginning of the pandemic

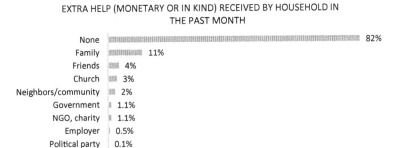

Fig. 4.5 Extra help received by households during COVID-19

COVID-19 Assistance Rare in Slum Communities

Few households in the study have benefited from extra assistance (monetary or in kind) since the pandemic struck Angola (Fig. 4.5). When received, most help came from family members (11%) or occasionally friends (4%). Over 80% of families, however, have not benefited from any extra material support. Despite the government's creation of special COVID-19 programs such as disbursement of free water or food baskets to the poorest of communities, and its reported acceleration of the monetary transfer program Kwenda,[13] only 1% of all participants received any kind of help from the government, suggesting that the small amount of government funds intended for slums (that are at a higher risk for COVID-19) trickle down with great difficulty. Studies from elsewhere in the Global South have shown similar difficulties when it comes to government help targeting and reaching the most vulnerable during the pandemic ([13], p. 11).

---

[13] https://www.unicef.org/angola/comunicados-de-imprensa/covid-19-coloca-em-risco-progresso-alcan%C3%A7ado-favor-da-crian%C3%A7a, accessed on June 13, 2021.

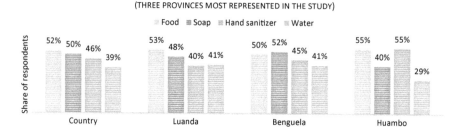

Fig. 4.6 Difficulty securing basic goods since the beginning of the pandemic

Difficulties Securing Basic Food and Services

Many households' budgets were already stretched thin since the time when an economic recession hit Angola in 2014–2015. The pandemic has further reduced the purchasing power of families and has automatically led to a reduced ability to buy goods of first necessity. Our research shows that since the start of the pandemic, one in two households have faced difficulty putting enough food on the table (Fig. 4.6). This food insecurity can lead to malnutrition and famine, a real risk for these households. This is especially true in the context of a severe drought that Angola is experiencing since 2019, making staple food rarer and further driving up its prices, especially in the driest provinces in the South-West. According to FAO predictions, at least 1 million Angolans will suffer food insecurity in 2021[14] [8]. Additionally, a significant proportion of households have been also struggling to buy soap (50%) and hand sanitizer (46%), and access to water in most provinces has been a challenge for nearly 40% of respondents across the country.[15] This observation is worrying as it jeopardizes hygiene practices in the middle of a pandemic.

Attitudes Toward COVID-19 Measures

The study also aimed to explore how various COVID-19 public health measures were received by slum communities. The results show a high awareness about schools closing and mask-wearing, while a third of respondents were not aware of the closing of commerce or public spaces (Fig. 4.7). A fourth did not know about the closing of markets. A large proportion (in all cases over 90%) supports mask wearing and creation of new ICU beds and does not feel harmed by the mask-wearing mandate. The closing of schools was supported by over 80% of households (Fig. 4.8), but one in two reported having faced difficulties tied to this measure, possibly related to childcare needs during the school closures (Fig. 4.9) A greater resistance can be observed

---

[14] Hunger Hotspots Report, March 2021, FAO—WFP.

[15] Huambo province traditionally enjoys a somewhat better access to water due to its rainy highland climate. Nevertheless, access to clean water remains a challenge.

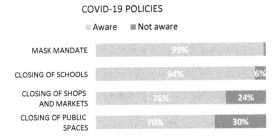

Fig. 4.7 Awareness of government COVID-19 policies

with regards to measures impacting the closing of public spaces and markets/shops, where 40% and 55% of households respectively expressed disagreement. Significantly, nearly 80% of households reported having been negatively affected by the closing of markets (Fig. 4.9). This is possibly related to the fact that not only do families buy food and convenience products in markets or on the street, but also because many of them earn income by selling in these spaces which have been targeted by the government bans. Our data above confirms that the negative socio-economic fallout of these restrictions has been felt by a large part of the slum households.

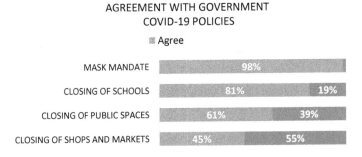

Fig. 4.8 Agreement with government COVID-19 policies

Fig. 4.9 Perceived impact of government COVID-19 policies

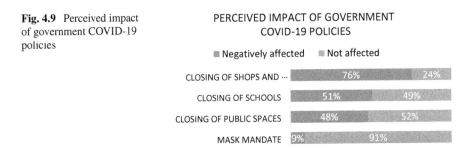

4.3 Pandemic Mitigation Strategies Through Participatory Slum Upgrading

4.3.1 The Potential of PSUP

Literature as well as experience on the ground reveal that no single program can represent a silver-bullet solution to the difficulties of slums—during and beyond pandemic times. A sustainable development of these informal areas can likely only occur through some combination of interventions at all levels (community, municipal, provincial, national, and international), across sectors (water and sanitation, waste, health, etc.) and actors (community, public, private, civil society etc.).

As exposed above, participatory slum upgrading can be a part of the puzzle in solving the "slum dilemma", and as recent studies suggest, institutionalizing this approach can help overcome its naturally incremental, piecemeal nature [9]. Countries such as South Africa, Tunisia or Mexico have benefited from a political will to implement national slum-upgrading policies, resulting in reduced slum-growth rates [16]. We hope that by sketching out a few practical participatory slum upgrading solutions here, we may open a space for discussion and fine-tuning of these approaches and contribute to a more inclusive, multi-stakeholder and sustainable slum governance that could become a model for urban authorities across Angola.

4.3.2 Participatory Slum Upgrading in Angola's Musseques in Pandemic Times

Before joining UN-Habitat PSUP, Development Workshop Angola had been involved in slum communities through projects related to water and sanitation, among others. As official COVID-19 prevention measures were coming into effect in Angola at the end of March 2020, DW decided to adapt its existing programs to the communities' needs in face of the pandemic. Organizing donations of biosafety material and funding awareness campaigns in local radio stations were some of the first steps that DW took in the slum communities during the pandemic. However, DW desired to go beyond these one-off activities to comply with the PSUP framework and to encourage a community-driven, sustainable pandemic response.

Promoting Inclusive Decision-Making

In promoting inclusive decision-making, it is advisable to always search for already existing bodies that have acquired a certain legitimacy in the eyes of the community and which could be leading PSUP activities, instead of creating new and untested organizational structures. In the Angolan context, these pre-existing institutions include the neighborhood traditional chiefs (*sobas*) or neighborhood commissions, church groups etc. Having the approval of an insider structure is an important requirement for local communities accepting and appropriating solutions introduced from outside, such as prevention strategies, COVID-19 vaccines, etc. [13]. A conscious effort needs to be made to include women in any of these groups, as women are often the overlooked yet crucial managers of community and family resources (water, food, finances) in day-to-day life. The youth, which now represents over half of Angola's population, is another demographic that must be given a seat at any table where decisions are made.

DW has also pushed for creating multi-stakeholder spaces where the exchange of ideas can take place (e.g. when dealing with water management, DW invited to one table neighborhood representatives, the municipal administration, the municipal directorates of energy and water, and other interested partners). Government can use these deliberative spaces to co-create COVID-19 policies adapted to the needs of the slum populations. For example, the government together with the communities can decide on a way of keeping the informal economy running even during the pandemic by adopting appropriate safety measures, as has been done in South Africa, Ghana or Senegal [19].

Capacity Building and Strengthening Resilience

Since the pandemic started, so-called "sanitation subgroups" constituted of local inhabitants were trained by DW to spread awareness in the communities and to lead social mobilization campaigns (Figs. 4.10 and 4.11). These local care committees [7] could be further trained to constitute emergency planning committees in order to coordinate response should the community face a major COVID-19 outbreak. DW has also organized training workshops for seamsters about the sewing of reusable masks which are a popular item sold in the local markets. Another activity led by DW has been the training of local women in artisanal soap-making using recycled cooking oil, aiming to promote hygiene, self-reliance, and sustainability.

Another way local populations can be empowered in the pandemic response while at the same time slum upgrading is through community management of water, sanitation, and waste [30]. DW has supported waterpoint caretakers through supplying biosafety materials and training them to be COVID-19 educators in the community. Training waterpoint technicians from within the community is equally promoting self-reliance and ownership of community resources while avoiding disruptions to the water supply. Informal waste-pickers and recyclers already indispensable in urban waste management can be engaged, and local clean-up teams (paid or volunteer) can be created to periodically dispose of loose trash in the community. This should be prioritized in and around waterways, near markets, and in other high-frequency areas to allow for better drainage and increased hygiene in public spaces.

Fig. 4.10 COVID-19 training for neighborhood activists in Mussulo, Cacuaco municipality, Luanda. November 2020

Fig. 4.11 COVID-19 prevention workshop at school N° 4059, Cacuaco municipality, Luanda. February 2021

Knowledge Co-production

As part of urban profiling, DW regularly organizes community mapping sessions in the slums where the locals are engaged to identify important locations frequented by the community (markets, taxi stops, playgrounds…), including those that could be used as isolation or vaccination facilities during the pandemic—a process similar to that of [20] in Kenya. This exercise has the potential to increase communities' preparedness for a potential disease outbreak, as well as to help in identifying facilities in need of upgrading.

Monitoring and Evaluation

Just like DW's profiling study ahead of launching an initiative, project evaluation must be continuous, and the local community must be listened to all along. Administering a periodical rapid evaluation questionnaire is one of the ways inhabitants can give feedback on a given PSUP initiative and signal any need for readjustment. DW has also been putting emphasis on evaluation meetings in communities to analyze the reach of the activity by autonomous community structures (neighborhood committees, associations of water consumers, etc.) but to also validate any gathered data—an exercise that has been made difficult by the pandemic restrictions on in-person meetings.

4.4 Conclusions

The objective of this paper was two-fold. Firstly, the authors aimed to stress the need for data collection as the first step ahead of participatory slum upgrading. We highlighted a few main results of a community study evaluating the impact of COVID-19 on Angolan slum households, conducted as part of the UN-Habitat-sponsored Participatory Slum Upgrading Programme (PSUP). Secondly, the article's aim was to show how PSUP can be relevant to slums during the current pandemic and beyond, rooted in knowledge co-production, inclusion, and sustainability.

The PSUP study has shown that the financial vulnerability of the slum households increased significantly, particularly due to reduced incomes, rising expenses, and little to no extra material assistance during the pandemic. This material hardship will likely trickle down to other areas such as food security, healthcare, education, crime, and mental health. Families welcomed certain government policies such as the promotion of mask-wearing and the temporary closing of schools, despite the difficulty of keeping children at home. But a large proportion of respondents showed a clear disagreement with policies restricting commerce and market sale. We argue that these policies not only increased hardship among the slum dwellers, but that they were also counter-productive since they oftentimes increased the risk of superspreading through overcrowding on the days sale was allowed. Interestingly, unemployment did not rise significantly since the onset of the pandemic, and no major differences in employment status between genders were observed. We assume this is due to the high involvement of the slum-dwellers in the informal economy, already high in pre-pandemic times. However, the high level of informal economic activity did not allow for more stability during the pandemic, as many of the abovementioned public health measures drastically curtailed informal employment opportunities for households.

Next, we sketched out how slum communities can become full-fledged agents of change alongside their governments and other stakeholders in mitigating the pandemic, all the while upgrading their living conditions. Capacity building and training (of artisans as well as of community educators), the use of community groups or committees, multi-party consultations, community mapping, and water, sanitation and waste management by the community all feature on the long, locally adaptable list of PSUP solutions. Regular follow-up surveys evaluating the impact of the PSUP activities will be crucial to adapting the project to the population's real needs and preferences.

We believe that the participatory slum upgrading activities we have presented and that are being implemented by DW since the onset of the pandemic can serve as a scalable roadmap of sorts to similar initiatives promoted by the government and other actors. In the short- and medium term they can be of use while dealing with the current health hazards, and in the long run they have the potential of increasing well-being and boosting preparedness of communities to face health—and other—emergencies.

Acknowledgements This article is based on a study made possible with the support of UN-Habitat Participatory Slum Upgrading Programme, with funding from the European Union. It was first presented at the 4th annual ICCCASU conference in Montréal, Canada in July 2021 (online).

References

1. Barros CP, Balsas CJ (2019) Luanda's slums: an overview based on poverty and gentrification. Urban Dev Issues 64(1):29–38
2. Bottan NL, Hoffman B, Vera-Cossio DA (2020) The unequal burden of the pandemic: why the fallout of Covid-19 hits the poor the hardest. Intra-American Development Bank https://doi.org/10.18235/0002834
3. Cain A (2014) African urban fantasies: past lessons and emerging realities. Environ Urban 26(2):561–567
4. Castro JC, Reschilian PR (2020) Metropolização e Planejamento Territorial Como Perspectiva de Desenvolvimento em Angola. Cadernos Metrópole 22(49):841–868
5. Connolly C, Keil R, Ali SH (2020) Extended urbanisation and the spatialities of infectious disease: demographic change, infrastructure and governance. Urban Stud 58(2):245–263
6. Croese S, Cirolia LR, Graham N (2016) Towards habitat III: confronting the disjuncture between global policy and local practice on Africa's 'challenge of slums.' Habitat Int 53:237–242
7. Dahab M, et al (2020) COVID-19 control in low-income settings and displaced populations: what can realistically be done? Confl Health 14(1)
8. FAO (2020) WFP early warning analysis of acute food insecurity hotspots

9. French M et al (2018) Institutionalizing participatory slum upgrading: a case study of urban co-production from Afghanistan, 2002–2016. Environ Urban 31(1):209–230
10. Friesen J, Pelz PF (2020) COVID-19 and slums: a pandemic highlights gaps in knowledge about urban poverty. JMIR Public Health Surveill 6(3)
11. Gill I, Schellekens P (2021) COVID-19 is a developing country pandemic. Brookings. https://www.brookings.edu/blog/future-development/2021/05/27/COVID-19-is-a-developing-country-pandemic/. Accessed 11 June 2021
12. González Pérez JM, Piñeira Mantiñán MJ (2020) La Ciudad Desigual en Palma (Mallorca): Geografía del confinamiento durante la pandemia de la COVID-19. Boletín de la Asociación de Geógrafos Españoles (87)
13. Gupte J, Mitlin D (2020) COVID-19: what is not being addressed. Environ Urban 33(1):211–228
14. INE (2021) Inquérito de Despesas, Receitas e Emprego em Angola (IDREA), 2019–2020
15. INE (2017) Anuário de estatísticas sociais, dados de 2011–2016
16. Jaitman L, Brakarz J (2013) Evaluation of slum upgrading programs: literature review and methodological approaches. https://publications.iadb.org/publications/english/document/Evaluation-of-Slum-Upgrading-Programs-Literature-Review-and-Methodological-Approaches.pdf. Accessed 14 June 2021
17. Leite C, et al (2019) Social urbanism in Latin America. Cases and instruments of planning, land policy and financing the city transformation with social inclusion. Future City Series, Springer Nature
18. Maggiola T (2020) Data favela Study: 80% of favela families are living on less than half of their pre-pandemic income. RioOnWatch. https://rioonwatch.org/?p=60404. Accessed 4 July 2021
19. Megersa K (2020) COVID-19 Crisis and the informal economy: executive summary. Women in informal employment: globalizing and organizing (WIEGO). https://opendocs.ids.ac.uk/opendocs/handle/20.500.12413/16766?show=full. Accessed 14 June 2021
20. Muungano wa Wanavijiji (2020) COVID-19 isolation centers: Muungano Alliance contributes to government guidelines. Muungano wa Wanavijiji. https://muunganoww.squarespace.com/browseblogs/2020/6/11/COVID-19-muungano-alliance-contributes-to-government-guidelines-on-isolation-centers. Accessed 23 June 2021
21. Osuagwu UL, Miner CA, Bhattarai D, Mashige KP, Oloruntoba R, Abu EK, Ekpenyong B, Chikasirimobi TG, Goson PC, Ovenseri-Ogbomo GO, Langsi R, Charwe DD, Ishaya T, Nwaeze O, Agho KE (2021) Misinformation about COVID-19 in sub-Saharan Africa: evidence from a cross-sectional survey. Health Secur 19(1):44–56
22. PMI (2020) FY 2020 Angola Malaria operational plan. https://d1u4sg1s9ptc4z.cloudfront.net/uploads/2021/03/fy-2020-angola-malaria-operational-plan.pdf. Accessed 5 July 2021
23. Purwanto E, Sugiri A, Novian R (2017) Determined slum upgrading: a challenge to participatory planning in Nanga Bulik, Central Kalimantan, Indonesia. Sustainability 9(7):1261
24. Rutherford S (2020) COVID-19 and the Hrishipara diarists: was April the cruellest month? Global Development Institute Blog. http://blog.gdi.manchester.ac.uk/COVID-19-and-the-hrishipara-diarists-april/. Accessed 15 June 2021
25. Sahasranaman A, Jensen HJ (2021) Spread of COVID-19 in urban neighbourhoods and slums of the developing world. J R Soc Interface 18(174):20200599
26. Sandoval V, Sarmiento JP (2020) A neglected issue: informal settlements, urban development, and disaster risk reduction in Latin America and the Caribbean. Disaster Prevention Manag 29(5):731–745. https://doi.org/10.1108/DPM-04-2020-0115
27. Taylor J (2020) How Dhaka's urban poor are dealing with COVID-19. International Institute for Environment and Development. https://www.iied.org/how-dhakas-urban-poor-are-dealing-COVID-19. Accessed 7 July 2021

28. Turok I (2021) Post-COVID cities can only be saved by reinvention and transformation. TimesLIVE. https://www.timeslive.co.za/sunday-times/opinion-and-analysis/2021-04-18-post-COVID-cities-can-only-be-saved-by-reinvention-and-transformation. Accessed 15 June 2021
29. UN-Habitat (Date unknown) Factsheet: participatory slum upgrading programme. https://unhabitat.org/factsheet-participatory-slum-upgrading-programme. Accessed 10 June 2021
30. UN-Habitat (2014) A practical guide to designing, planning, and executing citywide slum upgrading programmes. https://unhabitat.org/a-practical-guide-to-designing-planning-and-executing-citywide-slum-upgrading-programmes. Accessed 10 June 2021
31. UN-Habitat (2018) Documento Do Programa-País Habitat-Minoth Para O Desenvolvimento Urbano Sustentável De Angola. https://unhabitat.org/node/143365. Accessed 10 Oct 2021
32. United Nations (2020) UNSDG policy brief: COVID-19 in an urban world. United Nations. https://unsdg.un.org/resources/policy-brief-COVID-19-urban-world. Accessed 11 July 2021
33. World Bank and UNDP (2020) Confrontar as Consequências Socioeconómicas da COVID-19 em Angola. https://www.ao.undp.org/content/angola/pt/home/library/confrontar-as-consequencias-socioeconomicas-da-COVID-19-em-angol.html. Accessed 5 July 2021

Chapter 5
Corporate Social Responsibility (CSR) Response to COVID-19 in Africa: Towards Healthy Cities

Raynous Abbew Cudjoe

5.1 Introduction

5.1.1 Background

Corporate social responsibility (CSR) is a well-known aspect of company engagement within the local community [1–8]. However, there are many other options for CSR to benefit the world positively, such as human rights [9–13] and environmental protections [14–18]. Another meaningful way CSR can be used is to help support the local community in times of crisis [19–21], like the case of the COVID-19 pandemic.

The proper definition of CSR depends on whom it is expressed. Either way, it can be defined as a tool for enterprises to use to their benefit. The ethical view is to develop a sustainable and honest attitude towards the local community's problems or the world [22]. In current circumstances, the role of enterprises has changed to be more dominant and complicated. This research presents a case that shows a perfect implementation of CSR activities in urban development or regeneration of local communities in Africa. The selected case is not enough, yet it supplements the list of numerous enterprises that provide successful practices of CSR and is available in the company's CSR reports or the list of successful practices in CSR organizations. The selection of the specific African countries and the analysis of the theoretical framework of Urban CSR will enrich the current bibliography on this subject and define the current views on CSR and urban development, thereby highlighting the connection between urban development and CSR in the African context. This stresses the need for infrastructure and basic requirements for the communities, and for these multinationals tow operate in sustainable partnerships. Since it is not very easy to quantify the results of CSR activities with one globally accepted evaluated

R. A. Cudjoe (✉)
Shanghai University, Shanghai, China
e-mail: Raynous1@shu.edu.cn

© The Author(s), under exclusive license to Springer Nature Singapore Pte Ltd. 2023
L. Zhang et al. (eds.), *The City in an Era of Cascading Risks*, City Development: Issues and Best Practices, https://doi.org/10.1007/978-981-99-2050-1_5

system, it is crucial to present specific CSR examples with the possible effects on the communities based on the United Nations Sustainable Development Goals.

5.1.2 Urban CSR

CSR and urban development programs are often analyzed in developing countries where natural disasters or wars influence people's lives and the environment. In Europe, environmental activities, health, safety, and social activities are the most featured CSR activities in annual CSR reports. There are different CSR activities for emerging countries that face wars, earthquakes, hurricanes, etc. Still, this study only focuses on urban CSR cases that concern cities and neighbourhoods that face urban sprawl or just cities that seek to be competitive and develop resilient environments with the help of the private sector in COVID-19. Given specific conditions, the Urban climate responds to changes Landry. Therefore, the process of urban development is a positive outcome, and an answer to the opportunities and challenges present through these processes [23]. CSR plays its role in these changes by offering more opportunities for enterprises to develop sustainable environmental and social activities for cities. The relations between enterprises and communities are dynamic and either conflicting or harmonically. Enterprises require local communities to provide fair tax systems, a clean environment, infrastructures, educated and skilled citizens, and safety.

On the other hand, communities expect private sectors to organize and participate in social events, contribute to economic development by paying taxes, provide job opportunities, sustain their physical environment and protect the environment. This relationship between enterprises and cities means close partnership with public authorities or even governments. Corporate governance is a common definition of implementing policies, procedures, and reporting arrangements to ensure a company understands and manages its risks effectively [23].

According to Werna et al. [24], the main concept of community relations, citizenship, and governance are that companies should not focus only on their financial performance. They should endeavour to audit and report their social and environmental impacts and make substantial contributions to the communities in which they operate. The loopholes in the operation of these public instruments are filled with the philanthropy of the private sector (anticipating benefits ex tax relieves). They are a subject discussed among different parties as to the political turn of CSR [25].

5.1.3 Statement of Problem

COVID-19 has been hailed as one of the worst crises in years. It has led to unprecedented challenges for the entire world. It is only accurate to say that the effect on some countries, industries, and individuals is much worse than on others. Underdeveloped countries suffer significant consequences of this pandemic due to the airborne nature of the disease. Countries with poor city layouts and infrastructure are mostly affected. Although the negative impact of the pandemic is widely published, there exists a positive element, such as the commitment of communities and individuals to help one another, especially the vulnerable city dwellers. The private sector has taken positive action, yet there are persisting questions on how the private sector utilized its CSR activities to help the resilient city agenda within Africa viz-a-viz COVID-19. This study is built on a selected case from Huawei's CSR operation within the African context. The author chose four countries in the various African regions. Issues related to the CSR activities from the private sector in these countries were examined, and conclusions were drawn.

5.1.4 Research Aim and Objectives

This study aims to ascertain how CSR can be used to establish and maintain citizen health within African cities. The present study's objective is to analyze the role of CSR in building healthy cities during the phase of the COVID-19 pandemic.

5.1.5 Research Questions

Based on the preliminary information, the major question about the research is; how have CSR initiatives contributed proactively to the healthy and resilient city agenda in Africa during the COVID-19 scourage?

5.1.6 Significance of the Study

The significance of the present study lies in the examination of the versatility of CSR to assist in addressing different issues [26–32]. Traditionally, CSR has been used for engagement with employees and the community in which the company operates. However, many have engaged in humanitarian efforts [33–39]. The present study is significant in the academic and policy domains. The policy domain establishes the need for CSR initiatives to be aligned with the community's immediate needs. It also shows the need for an alliance between the private and public sectors to pursue a common agenda in the face of adversities. The present study advances the already existing CSR research in the academy and contributes to knowledge.

5.2 Literature Review

5.2.1 The Concept of CSR and Healthy Cities

A healthy city provides solutions to climate change problems in the immediate and external community [40]. The United Nations Sustainable Development Goals are collective agreements that enforce harmonious and sustainable development related to ecological, humanity, and economic aspects. In this regard, sustainable city development translates to creating livable city ecosystems. A livable city is a kind that provides ideal living conditions and sustainably provides space for city life.

The concept of CSR involves the responsibility of partnerships between government, private institutions, and stakeholders [41] in a proactive approach. The formation of a healthy and resilient city indirectly affects environmental sustainability [42]. So, there's the need to implement a control mechanism in designing healthy and resilient cities. On the other hand, there should be a limited funding ratio of CSR compensation in advertising media for CSR product providers. Thus, the corporate engagement model and design negotiation process are essential for integrated planning and designing in sustainable, healthy, and resilient cities.

According to McPhearson [43], establishing a private participation model through CSR in sustainable city development requires a comprehensive and integrated approach coupled with transparency and accountability in CSR funds.

City resilience has become a persisting issue and a focal area in landscape ecological research and urban problem research. In-depth analysis shows that socio-cultural, economic, and urban environmental problems also have implications for developing the concept of urban Resilience. In this regard, Kemperman and Timmerman [44] developed the idea of social Resilience by equating it with the ability to cope with climate change, natural disasters, and pandemics (COVID-19). Resilience, in this regard, defines a system's ability to overcome interference and rearrange while maintaining function, structure, identity, and feedback to normalize the already running system [45, 46].

The implementation of the healthy and urban resilience concept has evolved to include human social networks [47], adaptability to disaster recovery [48], security resilience [49], and even resistance to populating the COVID-19 pandemic. For this research, city resilience is defined as the City's ability to respond to internal and external risk pressures through absorption, adaptation, and transformation within existing basic structures and functions. The application of the definition lies in the City's ability to address urban problems related to climate change [50], pandemics (COVID-19), and disasters to take action to prevent and mitigate urban hazards [51]. In this sense, urban resilience as a process can be interpreted as an effort to increase the ability to absorb and respond to the effects of disasters and reorganize to overcome disruptions in achieving normal conditions after disaster stress or change [52]. A healthy and resilient city in the phase of COVID-19 should possess the ability to provide a livable environment for city dwellers during and after the COVID-19 scourge. Resilience in a system allows the system to adapt to change [53, 54].

In a nutshell, Resilience has three main functions: persistence, adaptability, and transformation ability, each of which integrates, collaborates and transcends from a local to a global scale.

Therefore, planning a healthy and resilient city requires evaluating the City's vulnerability, understanding of processes, procedures, interactions, and capacity building to develop consistent infrastructural components to achieve livable cities. Based on the description above, the research can lay bare the attributes of a healthy-resilient city.

5.2.2 Attributes of a Healthy City

With the rapid changes in the world, it is only suitable to trace the attributes of a healthy city from about a century back—down to the current standards of a healthy city. For this reason, the author referred to a report from Trevor Hancock [55] on "healthy cities in Europe". Hancock [56] wrote the first World Health Organization paper on Healthy Cities in 1988. This paper coined an inevitable definition in present-day discussions on healthy cities. In this research, the author gleaned 11 characteristics of a healthy city from the literature.

The definition:

A healthy city is constantly creating and improving the physical and social environments and strengthening the community resources, which enable people to mutually support each other in performing all the functions of life and achieving their maximum potential.

The key concept here is "continuity". It is emphasized to continue to do several things to become healthier. This implies the process of constant striving.

According to Hancock [56], The aim is broader than health: it is about functioning well in normal daily life and human development and potential. The focus is not just on the city itself but also the people there.

Hancock characterized healthy cities as one that meets the basic needs of city dwellers. It is a clean, safe, and high-quality physical environment with a stable, sustainable ecosystem in the phase of adversities such as COVID 19. A healthy-resilient urban space should possess a strong sense of community that is supportive and free from exploitation and encourage public participation to achieve city dwellers' collective health and well-being before and after any disaster. According to Hancock, one of the trademarks of a healthy-resilient city is its health status. A healthy city should possess a high health and low disease status. There should be high levels of appropriate public health and sick care services accessible to all and sundry to achieve this status. In desperate times like the COVID-19 pandemic, public health facilities serve as a tool to proactively assist in the control and revitalization process.

Furthermore, Hancock stated that a healthy-resilient city should be able to connect with the past, cultural and biological heritage while connecting with other groups of individuals for social purposes. In this regard, city dwellers are exposed to various experiences and resources with the possibility of multiple contacts, interactions and communication. One vital characteristic of a healthy city is a strong economy. A livable city should have a diverse, critical and innovative city economy that can adapt and bounce back after experiencing a harsh economic recession caused by COVID-19. A city compatible with and enhances the defined parameters and behaviours is also compatible with the UN SDG 11 (sustainable cities and communities) and is always resilient and ready to provide proactive solutions for city dwellers in the face of environmental, social and economic adversities.

5.2.3 Conceptual Framework of the Role of CSR in the Formation of Healthy-Resilient Cities

CSR in sustainable development focuses on the firm's responsibility to assume an organizational role in society, the environment, and business interests. To achieve excellence, companies take roles and responsibilities in developing the ecosystem. In the context of a healthy and resilient city program in the face of COVID-19, the study's conceptual framework demonstrates how firms liaise with city governments, community stakeholders, and pressure groups in providing city development funds and planning for the City's Resilience and revitalization during and after the COVID-19 pandemic. The government ensures that CSR initiatives and plans for the resilient and revitalization program meet the United nation sustainable development goal 11 (healthy cities and communities) (Fig. 5.1).

5.2.4 The Case of Huawei

Lateef and Akinsulore [57] noted, as did other studies in the prior subsection, that many disruptions have occurred due to COVID-19, leading to what is now considered the "new normal" for human activities. Africa has experienced a combined health and economic crisis, which has forced African state governments to resort to alternative fiscal policies to address the debilitating COVID-19 effects on city dwellers and economies. For African businesses, COVID-19 impacts have been just as disruptive as daily life due to the continuing trend of uncertainties regarding how business is conducted and sustainability [57]. Yet, there have been previously established strategic roles for corporations within African states, regardless of the state's economic power (but with the acknowledgement that some states have more economic power than others). Ultimately, per Lateef and Akinsulore, having sound corporate governance and CSR investments can assist in establishing, maintaining, and even improving the company's performance, which can be useful in addressing city needs and revitalization during and post COVID-19 pandemic. The case of Huawei posits an in-depth knowledge of how CSR activities by African companies

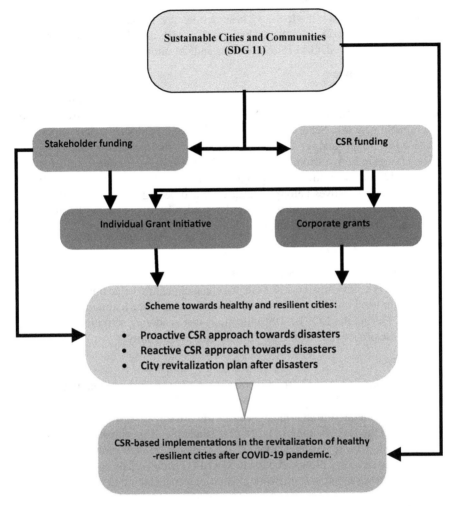

Fig. 5.1 Conceptual framework of the role of CSR in the formation of healthy-resilient city

should be aligned with the United Nations Sustainable Development Goals to build healthy and resilient cities in Africa.

West Africa—Nigeria

Huawei's CSR initiatives in Nigeria have mainly focused on proactive social contributions to improve living standards, support underprivileged groups, and reinforce social integration and cultural exchange and as a requirement of a healthy—resilient City, stated by Hancock. During the recent pandemic, the global tech giants acknowledged that students in Nigeria experienced various setbacks, especially in discharging school duties and social life. In response, Huawei donated computers and WiFi devices to students and teachers in the Ogun state of Nigeria to aid teaching and social lives amidst the challenges. These actions are in tandem with the United Nations Sustainable Development Goals 1, 3, 9, 10, and 13 and are relevant to applying the overall SDG 11 (Healthy cities and communities).

East Africa—Kenya

In Kenya, Huawei directed its CSR activities toward their product and service to contribute to the fight against COVID-19. Huawei's products ensured the digitization of a reliable network. Secure, high-speed, indoor and outdoor wireless application enhanced by routers and servers. In reaction to city lockdowns during the pandemic, these services in Kenya's communities helped city dwellers continue to have a normal work, study, and social life while keeping safe and healthy in their homes. Although proactive CSR initiatives before the pandemic had been geared towards this direction, Huawei's reactive response during the pandemic enhanced and improved the experience of city users. This aligns with the UN SDGs 8 and 11.

North Africa—Algeria

Meanwhile, in Algeria, the CSR activities, Huawei align with the UN SDG 4 and 11 by proactively cultivating local ICT talents, enhancing knowledge transfer and cooperating with local universities for cloud classrooms, distance learning, and research networks. During the COVID-19 pandemic, when academic activities were forced online, these proactive CSR initiatives were presented readily.

South Africa

In South Africa, it was realized that Huawei's CSR contributed socially to support education, nurtured local professionals, and drove digital transformation in 2020 during the pandemic through their "seed for the future" program. This contributes massively to the revitalization of cities post-pandemic. This enhances urbanization by contributing to quality education (SDG 4) and economic growth in the cities (SDG 9). These contributions are vital to SDG 11 (Healthy cities and communities). Figure 5.2 shows a simple framework of Huawei's contribution to healthy cities and communities.

Contrary to CSR activities by African companies providing only a reactive response to the COVID-19 pandemic, CSR activities by Huawei have contributed to the development of healthy cities based on the UN SDGs. Therefore it provides a proactive solution in the face of the COVID-19 pandemic. The case of Huawei is critical in the context of this study because it includes information that corporations can use in ascertaining their CSR policies for promoting societal needs and helping build healthy and resilient cities in Africa.

Fig. 5.2 Framework for Huawei's CSR contributions

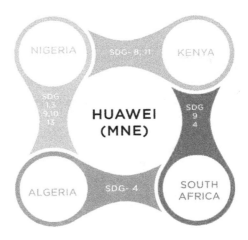

5.3 Methodology

The present study was conducted based on a five-stage systematic literature review, with the explicit goal of synthesizing current literature to draw new conclusions [58]. The research adopts a qualitative approach using a case study. The research methodology involves 4 African countries in which the selected Multinational enterprise operates. I picked a country from the north, south, east, and west of Africa to bring the whole of Africa in context. A single nation from a region might not be enough, but it creates an understanding of the wider regional population [67].

The data collection and analysis of the methodology are discussed conjointly because of the steps utilized in the systematic literature review. All steps are interrelated; thus, combining the data collection and analysis sections is the most effective approach.

Develop the research questions. This was done in the introduction based on the background, research aim, and research objectives.

Relevant data search. The inclusion criteria were developed for the present study. The search engine of choice was Google Scholar. As the topic is current, most of the literature is from existing sources. However, for discussions of CSR in African countries, references may be older, with the acknowledgement that seminal information may be slightly more ageing. Search terms were used in conjunction with the Boolean operator. The words were "COVID-19," "corporate social responsibility," "CSR," "Africa," "Initiatives", "Nigeria," "Kenya," "Algeria," and "South Africa".

Data eligibility. This stage was adopted to ensure the legibility of the literature sources using the Likert scale.

Afterwards, the author conducted a Data extraction process. Information relevant to the topic of interest was extracted for synthesizing and analysis.

Finally, data analysis. In this session, the data extracted in the prior step was analyzed to establish a new outcome [59, 60].

5.4 Discussions

5.4.1 Overview

This session discusses the social responsibilities in the selected African countries. The discussion addresses the impact of COVID-19 and the action of the public and private sectors to curb its effect on the countries down to individual communities. The individual cases in the selected countries reflect similar issues in the various African regions in the COVID-19 era. Different companies reacted in diverse ways using different CSR strategies concerning the various UN SDGs. Still, those in tandem with the concept of healthy and resilient city formation are selected for the study.

5.4.2 West Africa—Nigeria

The COVID-19 pandemic strike in Nigeria was very unprecedented. The country was already battling its oil market crisis while struggling with the country's dilapidated and inadequate infrastructure, leading the country to a deficit. These deficits affect the manufacturing sector as well as the tech sector. The major effect falls on digital services available for mobile money and online banking services. Given the current economic crisis and its impact on state finances in the face of COVID-19, the infrastructure deficit problem might worsen in the post-COVID era and diminish the potential of this sector. Also, the containment measures, which have led to business closures and movement restrictions, have brought private-sector consumption to a standstill indicating a 60% less value addendum. In response to this, the government has released its economic stimulus bill to mitigate the economic slump. The stimulus programme includes tax relief, cash transfers and the distribution of food packages for the poorest. However, the tech scene, which is grappling with a range of deficits, especially in energy and telecommunications infrastructure, has also received a wake-up call since the pandemic.

Before the pandemic, Cities in Nigeria already suffered a patchy and erratic energy supply. The available electricity capacity across the whole country is only about 4,000 MW out of 12,500 MW. This power deficiency causes constant power outages in major cities in Nigeria and results from the long-standing unavailable grid capacity and lack of regulation. Only around 45% of the population is connected to the grid, leaving the 55% remainders to solve their power outage problems. The same issue reflects in Nigeria's digital infrastructure. The total length of fibreglass cable infrastructure is around 64,400 km, roughly half that of Algeria or South Africa. Before the pandemic, the Nigerian Communication Commission addressed this issue, noting that implementing the 4G and 5G networks required doubling the number of telecommunications masts. The country is still a long way from achieving this goal. The COVID-19 pandemic has been more or less a reminder to policymakers on why the stated facilities are essential in building a resilient city [61].

The problems described above mainly affect the tech sector, which focuses on policy-making. Policymakers are pursuing a strategy to broaden the economic structure to mitigate and curb the influence of the resource economy. The ICT sector is not only essential for the fight against COVID-19, but it also generates around 13.8% of the country's GDP. The megatrend towards digitization can further support the healthy-resilient city plan in the revitalization process, especially since it has received considerable impetus from the coronavirus crisis. At the peak of the pandemic, when lockdowns were imposed, Jumia, Nigeria's largest online trader, reported year-on-year turnover growth of around 30% for the first quarter. The company and market observers attributed the rapid increase in sales figures to the coronavirus pandemic—Nigeria's growing and dynamic startup and technology sector address many existing problems in distributing goods and services. Making household deliveries during the lockdown, particularly in the middle and upper class, is only one example.

The startup and technology sector is believed to harbour great potential, particularly because of the dynamic population growth and the very young and, therefore, the tech-savvy environment in Nigeria. Still, the existing infrastructure problems are slowing down growth. This has been illustrated by the World Bank's current Business Index and the Center for Global Development. In response to the government's call, the private sectors, especially the technology sector, have Channeled their CSR activities to curb the power issues in the country as it is the first phase of the revitalization process.

The problems in electricity supply have encouraged the emergence of what is referred to as tech hubs or incubators. In addition to offering a reliable electricity supply, such facilities provide access to knowledge, capital and networking through other founders. Seventy incubators of this kind are currently active in Nigeria, more than any other African country. The notion is that digitization will thrive on successfully implementing a stable electricity supply. All aimed toward revitalizing healthy and resilient cities in Nigeria post-COVID-19.

5.4.3 East Africa—Kenya

According to the Kenya National Bureau of Statistics (KNBS), about 1.8 million people in Kenya lost their jobs between March and July 2020 as the pandemic transitioned into a major jobs crisis. Many Cooperatives struggled as others went into total shutdown. President Uhuru Kenyatta called on local governments and the private sector to help Kenyans cope during the coronavirus pandemic. The State Department of Cooperatives (SDC) quickly responded by forming the Kenya Cooperative Coronavirus Response Committee (CCRC), which comprises ten key stakeholder organizations, including national cooperative apex bodies.

The first CCRC initiative was designed to provide basic COVID-19 PPEs as the first line of defence against infection. This was done through cooperative societies, which distributed the kits within their communities to members and non-members alike. Private cooperatives have also contributed to tackling COVID-19 in their immediate cities and districts of operation. A survey by the US Agency for International CLEAR programme showed that Kenyan Cooperative businesses and their members have felt the effects of COVID-19 yet are actively working to meet the needs of their members and communities. The impact of the COVID-19 pandemic was great, such that; Kenya's cooperative sector reported a 30% decrease in revenues with some claimed losses up to 60%. The majority (70%) stated that social distancing restrictions had limited their operations, and over 30% of cooperative businesses sites closed markets or points of sale.

Businesses in the financial sectors responded to support efforts to contain and manage the financial crisis resulting from the pandemic. The insurance industry association issued an exemption notice covering coronavirus related causes as an act of goodwill. To aid learning for schools during the lockdown, The telecommunication giant Airtel Kenya offered free internet access on the Longhorn e-learning platform to enable students to continue studying. "… this will empower the students using its technology and ensure that students remain positively engaged in their learning process," MD Prasanta Das Sarma said. "This will help them continue learning while out of school and open up a new world of opportunities for their future," In the revitalization process after the pandemic is over.

5.4.4 Northern Africa—Algeria

As the largest African country and the 4th largest economy in the continent, the strike of the COVID-19 pandemic has exposed the vulnerabilities in the country's economic and infrastructure Resilience. To containment measures resulting from the 2020 COVID strike, the Algerian economy plunged by 5.5% [62]. Notwithstanding this, Summer 2021 brought another wave of infections, with case numbers surging from 500 to 1,500 a day amidst the country's health system close to buckling under the strain [63]. Political instability, societal unrest, corruption and a wildfire crisis have lent further uncertainty to the country's future. According to the government, digitalization stands to be the key to Algeria's development goals, economic diversification, financial inclusion and modernization of the banking and financial sector. These initiatives will be the major drivers to achieving resilient cities in Algeria, the implementation of the revitalization plan after the pandemic focuses on digitization. Before the COVID-19 crisis, 57% of Algerian adults and 71% of women did not have access to a transaction account. 16% of adults, and 11% of women, used digital payments. This made online purchases and transactions during the lockdown almost impossible for the masses and contradicted the Attributes of healthy cities [64]. The Algerian banking and financial sector rely on dated technological infrastructure [65]. Therefore it requires Fintech innovation to revolutionize its services, making them faster, cheaper, more accessible and easier to use.

Although Algeria's Fintech ecosystem is in its early stages of growth, there is a reason for optimism. The new presidential administration has been particularly progressive in its digital transformation efforts and has called on the private sector to aid it. Also, in an attempt to revitalize the economic space, the government launched the country's first Fund for Startups in October 2020 to introduce tax incentives for Startups in innovative and emerging technologies. In September 2020, amidst the COVID-19 crisis, the executive decree launched a national committee to classify Startups, creative projects and incubators. Algeria has recently improved the national digital infrastructure with the government's push to promote digital skills in education. This attracted Huawei's Seed innovation program, which cultivated local ICT talents and enhanced Knowledge transfer. The government initiatives and the CSR initiatives from the private sector during the pandemic in Algeria adopted an urgent response in dealing with economic and infrastructural deficits faced by the country. These initiatives stand not only as reactive responses but also as proactive responses in the long term towards the revitalization of healthy-resilient cities in Algeria after COVID-19 is long gone.

5.4.5 South Africa

As COVID-19 unravels the vulnerability of countries worldwide, African countries are challenged to respond to a global pandemic in poorer urban communities already facing multiple risks, and South Africa is no exception. Same as the climate crisis, COVID-19 is a risk multiplier adding to vulnerabilities from inadequate essential services, low incomes, unemployment, crime and domestic violence, and environmental and climate hazards, among many others. Citizens living in poorer communities are exposed to greater risks due to a lack of basic community infrastructure and amenities [66].

How do you wash your hands when you spend hours at communal taps to access the water supply? How do you practice social distance when you live in a single room with six other family members, and your neighbour lives less than three metres away? In response to this, the business community in South Africa reactively took a grip on the challenge in some of the largest and most densely-populated informal settlements in South Africa. The National Business Initiative (NBI) and Business for South Africa (B4SA) deployed a handwashing programme in the Ekurhuleni Municipality in Gauteng. Gauteng took the most hit by the coronavirus as the most populated province in South Africa. The government acted quickly to respond to the virus by galvanizing support from the private sector through CSR activities and grants.

In the words of Alex McNamara, NBI's Water Lead: "We learnt from our support with Day Zero in Cape Town that a crisis could quickly galvanize partnerships and build lasting bridges between business, government and civil society. In this instance, a coordinated business response is essential in the event of disasters. We now see such support from the private sectors playing out globally, with companies stepping up to play their part." The B4SA and NBI launched the emergency water response project to validate the overarching message WASH (water, sanitation and hygiene) against the coronavirus to create the first line of defence in local communities. The NBI and B4SA launched a fundraiser and technical assistance from businesses across the country in CSR grants. It also utilized a Solidarity Fund, which includes contributions from government, citizens, communities, and international donors to support South Africa's reactive efforts to respond to COVID-19 [66].

South Africa's Department of Water and Sanitation used the online risk and vulnerability mapping tool originally designed as a climate change adaptation tool in 2019 to identify 2,000 communities with no water services or inadequate handwashing facilities. A number of these communities are situated in Ekurhuleni, with over 100,000 people living in informal settlements.

NBI and B4SA, working with the City of Ekurhuleni, quickly assembled a team of engineers, scientists, water experts and project managers to design a foot-operated communal tap (as opposed to a hand-operated one), a clean handwashing facility at communal sites, and affordable handwashing units. The handwashing unit uses a two-litre plastic bottle design and is being initially trialled in 150 households in northern Johannesburg. Local private sector actors have designed these interventions, and implementation has begun in communities in less than two weeks. Water is also being supplied and stored in water tanks through the Department of Water and Sanitation work. Such initiatives demonstrate the critical need for proactive CSR activities toward healthy cities. They also hold potential lessons on CSR response to a crisis such as COVID-19 and how to support private sector action and innovations. Business groups like NBI and B4SA play a top facilitating role in rapid collective effort from the private sector, which enables quick disaster responses across different stakeholder groups. The impact of COVID-19 shows the importance of collaboration between the public and private sectors as a United front.

As stated in the abstract, the use of a single country in the four regions of Africa might not be enough, yet the evidence in the discussion reflects major problems and impacts from COVID 19. In west Africa, Nigeria suffered electricity supply problems in many cities, which directly contrasts the attributes of a healthy city. Algeria had digitization issues. South Africa showed a lack of basic amenities such as water and housing. Finally, Kenya suffered a financial crisis in the face of the pandemic. All these vulnerabilities in the major cities in these countries already existed. Still, the impact of the COVID-19 pandemic laid them bare and prompted the government and local governments and private multinationals to take reactive and proactive action towards them during the pandemic. The negative effect of the COVID-19 pandemic is widely published and discussed. Still, there's a positive element, such as the actions taken by corporations to build healthy and resilient cities in Africa.

5.5 Conclusions and Recommendations

Following the objectives of this chapter's discussion, several notes result from the discussion and conclusion and transcend to policy recommendations. Building a city with social, ecological, and esthetic functions is very important for realizing Urban

Resilience. Successful formation of quality healthy—resilient cities and communities will contribute to the health, social and economic growth of the city dwellers and the country. Access to a good community with standard housing and basic infrastructures like electricity, water, and internet can drive digitization in health, e-commerce, e-banking, and e-learning, making a simple daily life or even life in the phase of adversities such as COVID-19. It should be noted that the social and economic impact of the COVID-19 on a healthy—resilient city would be less severe than that of a city which lacks basic infrastructure and amenities. Collaboration between the government and the private sector's CSR in Africa is vital. The involvement of the private sector in the formation of a resilient city has a great prospect.

5.6 Future Works

For future researchers, it would be beneficial to conduct studies that consider the exact contribution of companies to alleviate problems associated with the COVID-19 pandemic. The situation is currently ongoing, so literature is scarce, but it will be possible to find this information soon. It is also recognized that many countries are just beginning to relax their social distancing restrictions, which makes it increasingly important to be aware of the risks associated with the reopening. However, by understanding the actions taken at each stage of the reopening process by companies to achieve CSR initiative goals, it may be possible for increased understanding of how to create healthy cities within African states and how they can be maintained effectively. By undertaking some of these studies, it would be possible to reduce complications from the limitations of the present study, such as the use of secondary data only. Other restrictions are selecting only a few African states and the limited range for data collection.

References

1. Ağan Y, Kuzey C, Acar M, Açıkgöz A (2016) The relationships between corporate social responsibility, environmental supplier development, and firm performance. J Clean Prod 112:1872–1881. https://doi.org/10.1016/j.jclepro.2014.08.090
2. Ararat M (2006) Corporate social responsibility across Middle East and North Africa. SSRN Electron J. https://doi.org/10.2139/ssrn.1015925
3. Arendt S, Brettel M (2010) Understanding the influence of corporate social responsibility on corporate identity, image, and firm performance. Manag Decis 48(10):1469–1492. https://doi.org/10.1108/00251741011090289
4. Betts A, Bloom L, Omata N (2014) Humanitarian innovation and refugee protection. Routledge
5. Hughey CJ, Sulkowski AJ (2012) More disclosure = better CSR reputation? An examination of CSR reputation leaders and laggards in the global oil and gas industry. J Acad Bus Econ 12(2):24–34
6. Mendibil K, Hernandez J, Espinach X, Garriga E, Macgregor S (2007) How can CSR practices lead to successful innovation in SMEs. Publication from the RESPONSE Project, pp 1–7

7. Park Y, Park Y, Hong PC, Yang S (2017) Clarity of CSR orientation and firm performance: case of Japanese SMEs. Benchmarking: Int J
8. Rupp DE, Mallory DB (2015) Corporate social responsibility: psychological, person-centric, and progressing. Annu Rev Organ Psychol Organ Behav 2(1):211–236
9. Emeseh E, Songi O (2014) CSR, human rights abuse and sustainability report accountability. Int J Law Manag
10. Mayer AE (2009) Human rights as a dimension of CSR: the blurred lines between legal and non-legal categories. J Bus Ethics 88(4):561–577
11. Ramasastry A (2015) Corporate social responsibility versus business and human rights: bridging the gap between responsibility and accountability. J Hum Rights 14(2):237–259
12. Wettstein F (2009) Beyond voluntariness, beyond CSR: making a case for human rights and justice. Bus Soc Rev 114(1):125–152
13. Wettstein F (2012) CSR and the debate on business and human rights: bridging the great divide. Bus Ethics Q 739–770
14. Edoho FM (2008) Oil transnational corporations: corporate social responsibility and environmental sustainability. Corp Soc Responsib Environ Manag 15(4):210–222
15. Lynes J, Andrachuk M (2008) Motivations for corporate social and environmental responsibility: a case study of Scandinavian Airlines. J Int Manag 14(4):377–390. https://doi.org/10.1016/J.INTMAN.2007.09.004
16. O'Connor A, Gronewold KL (2013) Black gold, green earth: an analysis of the petroleum industry's CSR environmental sustainability discourse. Manag Commun Q 27(2):210–236
17. Orlitzky M, Siegel DS, Waldman DA (2011) Strategic corporate social responsibility and environmental sustainability. Bus Soc 50(1):6–27
18. Ringov D, Zoilo M (2007) Corporate responsibility from a socio-institutional perspective: the impact of national culture on corporate social performance. Corp Gov 7(4):476–485
19. Kuznetsov A, Kuznetsova O, Warren R (2009) CSR and the legitimacy of business in transition economies: the case of Russia. Scand J Manag 25(1):37–45
20. Panwar R, Paul K, Nybakk E, Hansen E, Thompson D (2014) The legitimacy of CSR actions of publicly traded companies versus family-owned companies. J Bus Ethics 125(3):481–496
21. White CL, Alkandari K (2019) The influence of culture and infrastructure on CSR and country image: the case of Kuwait. Public Relat Rev 45(3):101783
22. Tsavdaridou M, Theodore M (2015) A theoretical framework on CSR and urban development. Munich Personal RePEc Archive 23(9):3–5
23. Roberts P, Hugh S (2000) Current challenges and future prospects. In: Urban regeneration a handbook. Sage Publications, London, pp 295–314
24. Werna E, Keivani R, Murphy D (2009) Corporate social responsibility and urban development. Lessons from the South. Palgrave Macmillan
25. Martin F (2011) Corporations as political and unpolitical actors. J Bus Ethics Organ Stud 16(2):12–21
26. Ali I, Jiménez-Zarco AI, Bicho M (2015) Using social media for CSR communication and engaging stakeholders. In: Crowther D (ed) Developments in corporate governance and responsibility, pp 165–185
27. Al-Masri S (2015) CSR, ethics and integrity in the Middle East enterprise space. In Entrepreneur. https://www.entrepreneur.com/article/250391
28. Belyaeva Z (2015) CSR in the Russian aviation industry: the winds of change. Strateg Dir 31(8):7–9. https://doi.org/10.1108/SD-06-2015-0094
29. Ben Youssef K, Leicht T, Pellicelli M, Kitchen PJ (2018) The importance of corporate social responsibility (CSR) for branding and business success in small and medium-sized enterprises (SME) in a business-to-distributor (B2D) context. J Strateg Mark 26(8):723–739. https://doi.org/10.1080/0965254X.2017.1384038
30. Bondy K, Moon J, Matten D (2012) An institution of corporate social responsibility (CSR) in multinational corporations (MNCs): form and implications. J Bus Ethics 111(2):281–299. https://doi.org/10.1007/s10551-012-1208-7

31. Boulouta I, Pitelis CN (2014) Who needs C.S.R.? The impact of corporate social responsibility on national competitiveness. J Bus Ethics 119(3):349–364. https://doi.org/10.1007/s10551-013-1633-2
32. Cernigoi A (2015) CSR in the Middle East: good business. In: Philanthropy age. http://www.philanthropyage.org/finance/csr-in-middle-east-good-business
33. Bealt J, Barrera JCF, Mansouri SA (2016) Collaborative relationships between logistics service providers and humanitarian organizations during disaster relief operations. J Humanit Logist Supply Chain Manag
34. CARR S (2019) Applying best practices from humanitarian aid to evaluate corporate social responsibility. In: The Oxford handbook of corporate social responsibility: psychological and organizational perspectives, p 48
35. Gjølberg M (2010) Varieties of corporate social responsibility (CSR): CSR meets the "Nordic Model." Regul Gov 4(2):203–229
36. Moggi S, Bonomi S, Ricciardi F (2018) Against food waste: CSR for the social and environmental impact through a network-based organizational model. Sustainability 10(10):3515
37. Rangan K, Chase LA, Karim S (2012) Why every company needs a CSR strategy and how to build it
38. Rupp DE, Skarlicki D, Shao R (2013) The psychology of corporate social responsibility and humanitarian work: a person-centric perspective. Ind Organ Psychol 6(4):361–368
39. Tarique K, Ahmed MU, Hossain DM, Momen MA (2017) Maqasid al-Shariah in CSR practices of the Islamic banks: a case study of IBBL. J Islam Econ Bank Financ 13(3):47–63
40. Wikantiyoso R, Suhartono T (2018) The role of CSR in the revitalization of urban open space for better sustainable urban development. Int Rev Spat Plan Sustain Dev 6(4):5–20
41. Dima J (2014) CSR in developing countries through an institutional lens. In: Corporate social responsibility and sustainability: emerging trends in developing economies, vol 8. Emerald Group Publishing Limited, pp 21–44
42. Juwito J, Wikantiyoso R, Tutuko P (2019) Kajian Persentase Ruang Terbuka Hijau pada Implementasi Revitalisasi Taman Kota Malang (Study of percentage of green open space in the implementation of Malang city park revitalization). Local Wisdom J Ilm Kaji Kearifan Lokal
43. McPhearson T, Andersson E, Elmqvist T, Frantzeskaki N (2015) Resilience of and through urban ecosystem services Ecosyst Serv 12
44. Kemperman A, Timmermans H (2014) Green spaces in the direct living environment and social contacts of the aging population. Landsc Urban Plan 129
45. Walker B, Holling CS, Carpenter SR, Kinzig A (2004) Resilience, adaptability and transformability in social-ecological systems. Ecol Soc 9(2)
46. Wardekker JA, de Jong A, Knoop JM, van der Sluijs JP (2010) Operationalizing a resilience approach to adapting an urban delta to uncertain climate changes. Technol Forecast Soc Chang 77(6):987–998
47. Adger WN (2000) Social and ecological resilience: are they related? Prog Hum Geogr 24(3):347–364
48. Buchanan M (2012) Disaster by design. Nat Phys 8(10):699
49. Godschalk DR (2003) Urban hazard mitigation: creating resilient cities. Nat Hazards Rev 4(3):136–143
50. Jabareen Y (2013) Planning the resilient city: concepts and strategies for coping with climate change and environmental risk. Cities 31:220–229
51. Balsells M, Barroca B, Amdal JR, Diab Y, Becue V, Serre D (2013) Analyzing urban resilience through alternative stormwater management options: application of the conceptual spatial decision support system model at the neighbourhood scale. Water Sci Technol 68(11):2448–2457
52. Davic RD, Welsh HH (2004) On the ecological roles of salamanders. Annu Rev Ecol Evol Syst 35(December):405–434
53. Meerow S, Newell JP (2015) Resilience and complexity: a bibliometric review and prospects for industrial ecology. J Ind Ecol 19(2):236–251

54. Meerow S, Newell JP, Stults M (2016) Defining urban resilience: a review. Landsc Urban Plan 147
55. Hancock T, Duhl L (1986) Promoting health in the urban context (WHO Healthy Cities Papers No. 1). FADL Publishers, Copenhagen
56. Hancock T, Duhl L (1988) Healthy cities: promoting health in the urban context (WHO Healthy Cities Paper #1). FADL, Copenhagen (Originally published in 1986 by WHO Europe, Copenhagen)
57. Lateef MA, Akinsulore AO (2021) COVID-19: implications for corporate governance and corporate social responsibility (CSR) in Africa. Beijing Law Rev 12:139
58. Hagen-Zanker J, Duvendack M, Mallett R, Slater R, Carpenter S, Tromme M (2012) Making systematic reviews work for international development research. Overseas Development Institute
59. Bamham C (2015) Quantitative and qualitative research. Int J Mark Res 57(6):837–854
60. Mack N, Woodsong C, MacQueen KM, Guest G, Namey E (2005) Qualitative research methods: a data collectors field guide
61. Heinemann T, Domnick C (2020) Nigeria's tech scene between coronavirus and infrastructure deficits. KfW research-focus on economics, no 305
62. The World Bank in Algeria. The World Bank, 14 June 2021. https://www.worldbank.org/en/country/algeria/overview#1
63. Hamdi R (2021) Algeria's health system is on the brink of implosion, as coronavirus wave hits. The Africa report, 04 August 2021. https://www.theafricareport.com/114474/algerias-health-system-is-on-the-brink-of-implosion-as-coronavirus-wave-hits/
64. Delort, D, Poupaert I (2021) How digital financial services can provide a path toward economic recovery in Algeria. World Bank Blogs, 22 February 2021. https://blogs.worldbank.org/arabvoices/how-digital-financial-services-can-provide-path-toward-economic-recovery-algeria
65. Fintech development in Algeria lags behind MENA counterparts. Fintechnews Africa, 02 June 2021. https://fintechnews.africa/39426/fintech-algeria/fintech-development-in-algeria-lags-behind-mena-counterparts/
66. McNamara L, McNamara A (2020) How businesses is responding to COVID-19 in South Africa's poorest urban communities. Climate and Development Knowledge Network
67. Landry C (2008) The creative city: a footprint for urban innovators. Earthscan Publications Ltd

Part II
Resilience and Risk in Contemporary Cities

Chapter 6
Southern Cities Between Rapid Urbanization and Increasing Need for Flood Mitigation Measures, The Case of Bujumbura, Burundi

Gamaliel Kubwarugira, Mohammed Mayoussi, and Yahia El Khalki

6.1 Introduction

Inundations are among the most common forms of risk. They are at the top of natural hazards as they cause 1/3 of natural hazards worldwide. They are also responsible of about 1/3 of deaths and 1/3 of injuries [1].

From a holistic perspective, it has been demonstrated that trends in the 20th and early twenty-first centuries are in line with substantial increase in the frequency and intensity of heavy rainfall producing recurrent flooding, especially flash flooding and urban flooding [2].

Then, those urban flooding are happening in a context of rapid urbanization characterized by proliferation of informal human settlements resulting in dwellings that do not meet construction standards [3].

Through concentration of population and infrastructure, urbanization significantly increases the predisposition of human settlements to being affected by natural hazards [4]. The social, political and economic environment also plays a major role in generating and worsening natural hazards [5].

Hence, stemming from actual trends of increasing urbanization and important climate variability, there is a need for deeper understanding of urban infrastructures vulnerability [6]. In this way, actions could be taken: among others implementing

---

G. Kubwarugira (✉)
Department of Applied Sciences, Burundi Superior Teachers' School, Bujumbura, Burundi
e-mail: kubiel@yahoo.fr

M. Mayoussi · Y. El Khalki
Department of Geography, Sultan Moulay Slimane University of Beni Mellal, Beni-Mellal, Morocco
e-mail: y.elkhalki@usms.ma

© The Author(s), under exclusive license to Springer Nature Singapore Pte Ltd. 2023
L. Zhang et al. (eds.), *The City in an Era of Cascading Risks*, City Development: Issues and Best Practices, https://doi.org/10.1007/978-981-99-2050-1_6

risk-sensitive Land Use and development policies aimed at protecting poor people especially in African cities [7].

In developing countries, although the lack of baseline data (that can be used to delineate flood-prone areas) leads to the use of simplistic approaches to this effect, it remains crucial to identify, quantify and communicate efficiently the extent of the flood risk [3]. Those analyses can be performed in different ways.

In Pakistan, the two-dimensional simulation rainfall–runoff model has been applied to analyzing flood extents. It was then possible to delineate flood extents and point out hotspots where the risk is greater [8].

In Tanzania, using the L-moment method diagrams, statistical distributions were established for various regions to provide engineers with sufficient information about flood estimates to make it possible to judge if a given site is suitable for receiving constructions. Results showed that it was possible to determine the extents of floodplain and also design accurately flood control structures [9].

In Rwanda, effects of flood in exposed areas have been analyzed for reduction of their magnitude. It has been shown that proposing the design of hydraulic structure is a good way to mitigate inundations in flood prone areas, especially wetlands like Nyabugogo in Kigali [10]. In the same context (Rwanda), another research based on Weighting Linear Combination (WLC) showed that vulnerability is quickly extending in Rwanda countrywide and Kigali city. Appropriate vulnerability lessening and adaptation mechanisms were proposed as well [11].

In the case of Burundi, the National Development Plan [12] considers natural hazards as an issue of security and national defense (see Axis 14 of the plan). The Government has therefore put a particular emphasis on the need to develop a Geographical Information System on the risks and make available special data on high-risk environments as a monitoring and evaluation indicator.

In our study, we analyzed the trends in buildings exposed to flooding in reference to historical flood episodes that occurred in Bujumbura City. We considered the rise of Lake Tanganyika level in 1878 and 1964 and the exceptional precipitation height recorded in 2014 that resulted in overflow of rivers cutting across Bujumbura City.

Our objective was to estimate the number of housing units that could be actually affected by flooding if there occurred episodes comparable to the recorded rise of Lake Tanganyika level or overflow of streams cutting through Bujumbura. Comparing exposition degree recorded in the past to actual prediction of housing and other goods exposition to flooding was then expected to allow us to draw some conclusions pointing out the urgent necessity to integrate measures related to flood risk mitigation in urbanization by-laws.

6.2 Methodology

In order to map the exposure of buildings to flooding in Bujumbura, we considered the state of land-use in 2011 according to the Constructions Identification Plan in Bujumbura urban area. As for the urban perimeter, it was possible for us to choose from three references: the administrative boundaries defined by Bujumbura City Council in 1992, the urban perimeter defined by Geographic Institute of Burundi in 2013 [13], as well as the boundary established by Bujumbura Master Plan in 2014 [14].

The 2014 Master Plan considers as Bujumbura urban area a sector consisting of 10,461 ha, while the official administrative perimeter defined in 1992 corresponds to an area of 11,458 ha. In our study, we therefore considered an urban area consisting of 13,012 ha as determined by the IGEBU [13].

Furthermore for analysis purposes, we took into account the fact that the risk of flooding faced by the Bujumbura City stems mainly from two sources: Lake Tanganyika which delimits Bujumbura to the West and rivers cutting through the City [15].

Hence, mapping of floods from Lake Tanganyika referred to two major rising sea level events. The first episode happened in April 1878. For this episode, water from the lake reached an exceptional altitude of 783.60 m (Usumbura[1] 1964), on one hand. On the other hand, we have the flood of May 16, 1964, associated with an increase in the water level up to 777.08 m.

For the two cases of rising of the Lake Tanganyika level, we carried out the corresponding mapping by identifying the floodable coastal dryers by a geospatial query based on two criteria: altitude below the water rating and connection to the shore of the lake.

As for the risk that derives from overflowing rivers through Bujumbura City, we evaluated it by taking as reference the exceptional precipitation of February 09, 2014 which caused important devastation especially in the northern neighborhoods of the city.

Mapping of this last flood risk extents, which is more complex than the first, has already been carried out using a method combining hydrological modelling of water flows and flood simulation based on runoff flows and terrain morphology. The Rational Method was applied to this end [16]. The work presented here has been to move the model forward and identify buildings in flood areas that could be affected by river overflows.

To this end, we felt it essential to extend the study area beyond the urban perimeter because most of the watersheds that feed the rivers crossing the city extend out of it, in the metropolitan area.

Overall, we considered 7 watersheds. Extent of each of them is totally or partially included in the urbanized perimeter or in the extensions planned by the 2014 Master Plan to the North and South of Bujumbura. Delineation of the watersheds (Fig. 6.1) to be considered in our analysis was performed with spatial analysis tools available in the QGIS software version 2.12.2.

Since our study area has already been subject to analysis of flood risk from river overflows and surface water runoff [15], we compared the results of the model implemented by ourselves to those from previous research in order to validate our model or modify it if necessary.

---

[1] Cited by Sindayihebura [15].

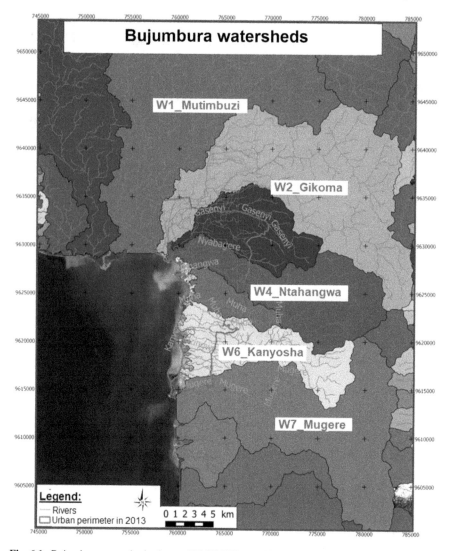

Fig. 6.1 Bujumbura watersheds. *Source* USGS [17]; modified by authors

This allowed us to confirm at the first time the geometric and dimensional characteristics of the delineated basins and at the second time, the results obtained for peak flows calculated to be input during the flood simulation process.

Table 6.1 below shows the study watersheds and their geometric characteristics that were utilized for calculating the concentration times and runoff flows.

Figure 6.1 illustrates delineation of watersheds in the study area.

Table 6.1 Watershed characteristics

Watershed	Length of watershed L (km)	Area; A (km^2)	Length of watershed L (km)	Area; A (km^2)
W1_Mutimbuzi	120.89	338.50	–	–
W2_Gikoma	128.59	270.71	–	–
W3_Nyabagere	39.47	77.28	71.09	123.55
W4_Ntahangwa	71.10	125.06	28.02	25.83
W5_Muha	28.01	28.68	–	–
W6_Kanyosha	46.45	90.14	–	–
W7_Mugere	99.82	215.79	–	–

Source Authors' calculations and Sindayihebura [15]

Table 6.2 Flood extent and number of buildings affected by historic flooding in bujumbura

Flood episode	Area (ha)	Area (%)	Number of buildings	Buildings (%)
April 1878 floods (Usumbura[2] 1964)	2912.94	22	6188	6
Floods of May 1964	1079.35	8	1655	2
Floods of February 2014	1691.64	13	18,747	19
Bujumbura (Overalls)	13,012.08	–	99,605	–

Source Authors' calculations

For the cases of Ntahangwa and Muha where we have data for comparison, the results of the calculation of geometric characteristics are very similar to the characteristics of the watersheds determined before us by Sindayihebura [15]. This proves that our delineation of watersheds is consistent.

6.3 Results

After mapping the flood areas for the three flood events selected for our study, we calculated extents of flooding and the number of buildings affected by the floods in each of the three cases. Table 6.2 summarizes calculations of the affected areas and buildings by comparing them to the total area and number of buildings in Bujumbura City. Figures 6.2, 6.3 and 6.4 show the location of the flooded areas, the extent of the floods and the affected buildings for the flood events of 1878, 1964 and 2014, respectively.

---

[2] Cited by Sindayihebura [15].

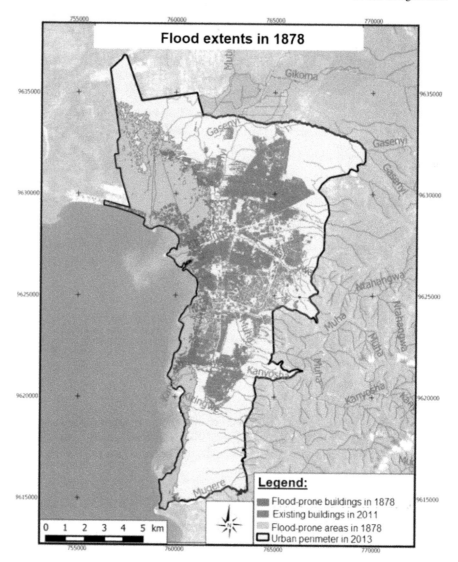

Fig. 6.2 Extent of flooding in bujumbura in 1878 [15]. *Source* IGEBU [13], and Sindayihebura [15]; modified by authors

6.4 Discussions

As explained by Sindayihebura [15], there is a rapidly increasing number of buildings likely to be flooded as a result of rising Lake Tanganyika level and overflow of rivers crossing through Bujumbura City. This is very logical given that many neighborhoods are developing by progressive densification. Moreover, many households and

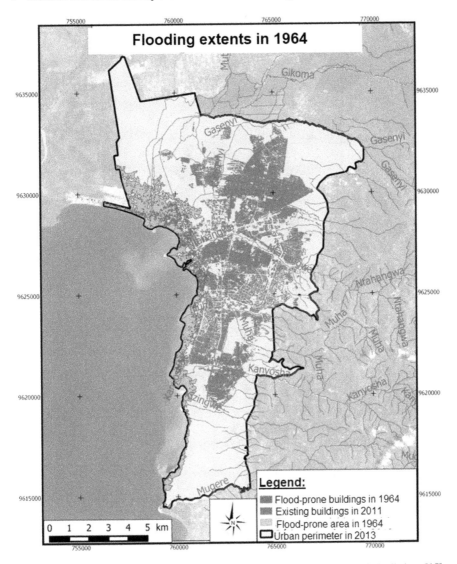

Fig. 6.3 Extent of flooding in bujumbura in 1964 [15]. *Source* IGEBU [13], and Sindayihebura [15]; modified by authors

companies are illegally erecting a huge number of buildings in the Lake Tanganyika perimeter and along different rivers, most of which are spaces highly exposed to flooding.

Thus, on the one hand, our results show that the number of buildings exposed to flooding is now far higher than the number determined by Sindayihebura [15]. Our results sometimes range from simple to double the number of buildings calculated

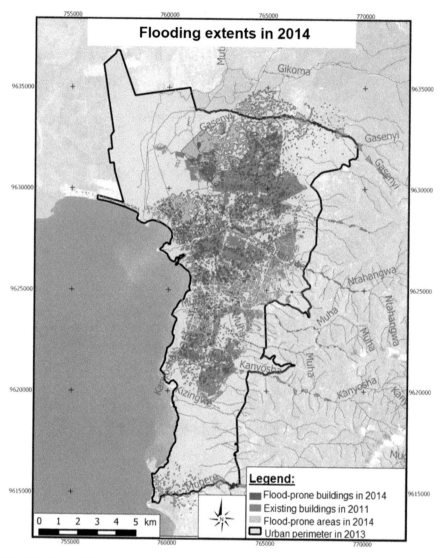

Fig. 6.4 Extent of flooding in bujumbura in 2014. *Source* IGEBU [13], and Sindayihebura [15]; modified by authors

by this author on the basis of the state of Land Use in 1994 for the entire city of Bujumbura and in 2002 for the neighborhood of Kibenga (Southern Bujumbura).

On the other hand, if we compare the number of buildings exposed according to the two flood episodes of 1878 and 1964, we find that our results seem to be minimized in the case of 1878 since the ratio between the two episodes is 3 times higher in Sindayihebura [15] than in our results.

In fact, we explain this significant gap as evidence that corroborates the above mentioned observations about construction within the logically dedicated perimeter of Lake Tanganyika, considered vacant land by some residents. Furthermore, the extent of the flooded perimeter in 1878 is likely to affect less buildings compared to the area flooded in 1964 because the areas located in the sea level difference of 6.52 m separating the two flood episodes, are already densified between the two dates and there are fewer new constructions, unlike the fringe along the shore of Lake Tanganyika, which is common to both episodes.

These findings are confirmed by the flooding that has persisted since the end of 2020 until 2021, following slight increases in the level of Lake Tanganyika. The intensification of damage after each variation in the water level of the lake is an indicator clearly showing that many residential neighborhoods as well as hostels and leisure establishments have been developed without the urban planning authority having taken appropriate measures, capable to ensuring security of these constructions.

As for comparison between the event of rise of Lake Tanganyika level in 1878 and the recent overflows of streams crossing the city (09 February 2014), there is a noticeable difference in the number of exposed houses. While the floods of 2014 accounted for only one third of the area flooded in 1878, the number of buildings affected is three times higher than the number of housing units affected by the 1878 event.

This difference is even more noticeable when comparing the proportions of the flooded area and buildings in 2014 compared to the city as a whole. We note that this flood event affected only 13% of the urban area of Bujumbura, but it damaged 19% of the buildings.

From the above, it is clear that floods derived from rivers are more dangerous than those from Lake Tanganyika because they cross the heart of densely populated neighborhoods. Moreover, it is quite understandable that with the development scenario adopted by the 2014 Master Plan which directs short- and medium-term urban growth towards the Northern sector of Bujumbura, exposure to flooding will be exacerbated.

Moreover, this kind of situation where the city faces the challenge of overexposure to the risk of flooding in connection with its growth is not a reality specific to Bujumbura. Similar problems have been identified for example in the city of Beni Mellal (Morocco), which is increasingly confronted with recurrent flood episodes especially during the months of September, November and December. In this city, current fronts of urbanization encompass traditionally flood-prone areas [18].

Finally, even if the lands located in the Northern part of Bujumbura City and the coastline of Lake Tanganyika form sites that can be easily built in relation to regularity of terrain altitude, it is important that appropriate measures be taken to ensure that future constructions to erect in these locations are protected against flooding.

6.5 Conclusions

As the phenomenon of spontaneous urbanization becomes widespread, many studies on flood exposure in unplanned human settlements are needed because spontaneous urbanization is no longer simply a reality for marginal groups of the urban population, but a dynamic on which relies on the urbanization process of many cities, especially in developing countries. This has for example been proven in the case of the city of Beni Mellal, Morocco [19].

From this spontaneous growth of cities, a kind of urbanism of regularization follows, most often leading to depletion of agricultural land, which is the most fertile [20]. Moreover, in the case of occupation of high-risk sites, operations dedicated to regularization are in no way a remedy, because even if there is a contribution of urban equipment and infrastructure, actions to protect spontaneous neighborhoods from flooding are rarely implemented. In these circumstances, local government intervention is most often limited to a simple crisis management [21].

It is therefore clear that, in addition to crisis management in spontaneous urbanization districts, a more sustainable alternative would be preventive in nature, based on assessment of the degree of flood exposure associated with different sites of the urban perimeter and the surrounding areas where the expansion is likely to occur, as shown in this paper.

This can be performed using spatial analysis methods such as the one proposed here because they make it possible to delineate areas prone to flooding, so that residents and urban growth managers are aware of the risk related to construction in the various sectors constituting the urbanization front, so that necessary accompanying measures should be taken.

References

1. Askew, A.J. (1999) Water in the International Decade for Natural Disaster Reduction. In Leavesley, G.H., et al., Eds., Destructive Water: Water-Caused Natural Disaster, Their Abatement and Control, IAHS Publication.
2. Zbigniew W et al (2014) Flood risk and climate change: global and regional perspectives. Hydrol Sci J 59(1):1–28
3. De Risi R et al (2013) Flood risk assessment for informal settlements. Nat Hazards 69(1):1003–1032
4. ActionAid (2016) Strengthening urban resilience in African cities: Understanding and addressing urban risk. ActionAid International, Johannesburg
5. Blaikie P et al (1994) At risk: natural hazards, people's vulnerabil-ity and disasters. Routledge, New York
6. Gannon KE et al (2018) Business experience of floods and drought-related water and electricity supply disruption in three cities in sub-Saharan Africa during the 2015/2016 El Niño. Glob Sustain 1(e14):1–15
7. Winsemius HC et al (2018) (2018) 'Disaster risk, climate change, and poverty: assessing the global exposure of poor people to floods and droughts.' Environ Dev Econ 23:328–348
8. Siddiqui MJ et al (2018) Rainfall–runoff, flood inundation and sensitivity analysis of the 2014 Pakistan flood in the Jhelum and Chenab river basin. Hydrol Sci J 63(13–14):1976–1997

9. Muhara G (2001) Selection of flood frequency model in tanzania using L-moments and the region of influence approach. In: 2nd WARFSA/WaterNet symposium: integrated water resources management: theory, practice, cases. Cape Town
10. Munyaneza O, Nzeyimana YK, Wali UG (2014) Hydraulic structures design for flood control in the Nyabugogo Wetland, Rwanda. Nile Basin Water Sci Eng J 6(2):2013
11. Nahayo L et al (2019) Climate change vulnerability in Rwanda, east Africa. Int J Geogr Geol 8(1):1–9
12. Burundi Government (2018) Plan national de développement du Bu-rundi: PND BURUNDI 2018–2027. Unedited
13. Carte topographique au 1 : 25 000 - Bujumbura et ses environs (2013), IGEBU, SA-35-XXIV
14. Burundi Government-Ministry of Transports, Publics Works and Equipement (2014) Schéma Directeur d'Aménagement et d'Urbanisme de la ville de Bujumbura à l'horizon 2025. Bujumbura. 250 p
15. Sindayihebura B (2005) De l'Imbo au Mirwa. Dynamique de l'occu-pation du sol, croissance urbaine et risques naturels dans la région de Bujumbura (Burundi). PhD thesis, University of Toulouse II, France
16. Kubwarugira G, Mayoussi M, El Khalki Y (2019) Assessing flood exposure in informal districts: a case study of Bujumbura, Burun-di. J Appl Water Eng Res
17. Landsat images for Bujumbura, Landsat 4–5 TM C2 L2 (2016), United States Geological Survey
18. El Hannani M et al (2014) Risques hydrologiques et d'effondrement dans le Tadla. Les dynamiques paysagères en question à travers le cas de la ville de Beni Mellal. In: Ballouche et Taïbi (eds)
19. El Khalki Y et al (2007) Processus d'urbanisation et accroissement des risques à Beni Mellal (Tadla-Azilal, Maroc): apports des SIG et de la télédétection. Mosella XXX 1–4:147–161
20. Balanche F (2009) L'habitat illégal dans l'agglomération de Damas et les carences de l'Etat. Revue Géographique de l'Est 49/4
21. Wilma SN (2007) Flood risk in unplanned settlements in Lusaka. Environ Urban 19(2):539–551. https://doi.org/10.1177/0956247807082835

Chapter 7
The Heritagisation of Social Housing, a Model of Ordinary Urban Resilience?

Géraldine Djament

7.1 Introduction: For a Triple Shift in the Approach to Urban Resilience

A social geography perspective allows a different approach to the issue of urban resilience. The intersection of three usually separate approaches offers a triple shift from the official stance…[1]

(1) **To the resilience of the ordinary city** [1, 2], **of all city dwellers in metropolisation.**[2]

Although the study cases are part of the Paris metropolis, located in a northern country and recognized as one of the main global [3] and world cities [4], they remain peripheral and working-class neighborhoods.

(2) **To a critical perspective on urban "sustainability" and urban "resilience".**

Urban resilience(s) [5] is a very polysemic concept [6] indicating at least a process of recovery after a disturbance. These concepts will be approached critically, regarding

---

[1] This paper is based on my communication at the round table organized by Patricia Zander (Strasburg university, France) "Towards otherwise resilient cities?/Vers des villes autrement résilientes?" Saturday, July 31, 2021, within the framework of the thematic area "Towards More Resilient Cities Worldwide" coordinated by John Zacharias (Peking University, China) and Inês Macamo Raimundo (Eduardo Mondlane University, Mozambique).

[2] Friend whom I warmly thank for her help.

G. Djament (✉)
Strasbourg University, Strasbourg, France
e-mail: djament@unistra.fr

not only the environmental[3] aspects but especially the social issues, and raising the fundamental question: whom does the resilience process benefit? Is it accompanied by social reproduction [11] or even social exclusion, or does it represent a popular urban resilience?

(3) **To an original approach from the point of view of the heritagization process,** "process of collecting and valorising whereby a social group takes an object—in the broad sense of the word—out of the ordinary scheme of things and elevates it to the status of emblem of its identity in time" (translated from Micoud 2005).

Although rarely associated, the concepts of heritagization and of resilience, in its etymological sense of rebounding backwards, present the same temporal structure consisting in drawing from the past to project towards the future and transmit. This chapter, anchored in a social geography of heritagization [12, 13] which focuses on the ways of inhabiting heritage [13], also hypothesizes **a scenario of urban resilience through heritagization**, based on the study of the recent and paradoxical heritagization of social housing [14–16].

The two popular field research sites that support this theoretical hypothesis are located in the northern suburbs of Paris: the garden suburb of Stains is close to the northern limits of the Seine-Saint-Denis, while the large complex of Maladrerie in Aubervilliers almost borders the capital city. Approached through semi-directional interviews and participatory observation during visits or animations conducted from 2013 to 2022, these two neighborhoods are part of the intercommunality of Plaine Commune, built in the early 2000s by the so-called reformist communists by pooling urban regeneration operations between Saint-Denis and Aubervilliers (Fig. 7.1).

It now has more than 435,310 inhabitants in 9 municipalities[4] and has since January 1, 2016, become a territorial public establishment of Greater Paris. This territory, historically marked by industry, birthplace of French social housing [17], which houses today 46% of the inhabitants, according to the French institute for statistics, is an instance of an early and active heritagization and tourist invention in "banlieue" (suburbs) or even of the "banlieue" itself [18], that is to say claiming a "banlieue" identity as such.

From these two experimental case studies, we will seek to show **how heritagization can provide an alternative and popular scenario of urban resilience**.

---

[3] This name replaces in 1950 in France the previous "Habitations à Bon Marché" and marks the transition to the "general model" of social housing.

Let us recall the international role of the issue of resilience after a non « natural » hazard, the attacks of September 11, 2001 in New York, in the dissemination of this notion in the field of social sciences [7]. In France, the concept of urban resilience has been confronted with that of urban perenniality [8] and is now not only used in the field of risk geography [9], but also in other fields: urban geography (cf. P. Zander (Strasbourg University, France) in the same round table) and/or tourism, economic geography (cf. K. Marius (Bordeaux University, France) in the same round table), sometimes with a question about "alternative practices for urban resilience" in the face of the economic crisis linked to deindustrialization [10].

[4] Aubervilliers, Épinay-sur-Seine, La Courneuve, Île Saint-Denis, Pierrefitte, Saint-Denis, Saint-Ouen, Stains, Villetaneuse.

7 The Heritagisation of Social Housing, a Model of Ordinary Urban ...

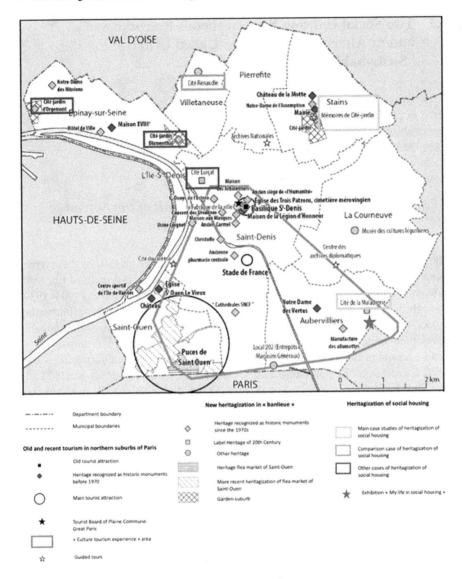

Fig. 7.1 Map of the heritage and tourism of Plaine Commune and localization of case studies (conceived by G. Djament, designed by V. Lahaye, 2017, retouched by G. Djament, 2022)

7.2 Two Social Housing Neighborhoods Precursors of and/or Alternatives to the So-Called Urban "Sustainability"

These two neighborhoods, which are largely made up of social[5] and green housing, serve to put into historical perspective and relativize the current stance on "urban sustainability": in urban planning, the link between city and nature has been topical since the late nineteenth century.

7.2.1 The Garden Suburb of Stains, a Result of the Spread in France of the Garden City Movement

The garden suburb of Stains, an adaptation in the Paris region [19] of the international urban planning movement initiated by the British reformist socialist Ebenezer Howard [20] and introduced in France by Georges Benoît-Lévy (1903–1939) throught his book *La cité-jardin* (1904) and the Social Museum, "is an interesting model, which is coming back into fashion with is the current emphasis on sustainable development and eco-neighborhoods", states the official in charge of tourism.

On the site of the former castle of Stains, which was destroyed during the war of 1870, the architects Eugène Gonnot and Georges Albenque created between 1921 and 1933 for the Service of low-cost residences (Office d'Habitations à Bon Marché) of the Seine Department created in 1915 and led by the reformist socialist Henri Sellier (1883–1944) [21] a garden suburb composed of 1600 housing units in Art Deco style [22]. On 28 ha are mingled small apartment buildings and detached houses, various equipments and ubiquitous vegetation: front gardens (ornamental) and back gardens (sometimes vegetable plots), green spaces at the heart of the blocks, English-style closes, greening of urban composition. The initial inhabitants were workers from the factories of Saint-Denis or La Courneuve (Fig. 7.2).

7.2.2 The Maladrerie in Aubervilliers, a Combinatory Architecture

Much more recently, the large complex of Maladrerie constitutes, on 8 ha, a variation of "combinatory" architecture. Between 1975 and 1984, Renée Gailhoustet (1929–2023) [23], recipient of several national and international prizes for architecture since 2014, oversaw the construction, for the Aubervilliers housing Office (Office d'HLM), and at the request of the mayor of the city, of a new district, applying the

---

[5] Even though Maladrerie has had 51 condominiums since the beginning, located at one end of the neighbourhood.

7 The Heritagisation of Social Housing, a Model of Ordinary Urban ...

Fig. 7.2 The greening of the garden suburb of Stains *Source* G. Djament, like all photographs in this chapter

principles developed with Jean Renaudie (1925–1981) for the redevelopment of the city center of Ivry sur Seine [24]: star-shaped buildings with garden terraces, large green spaces, urban planning alternatives to the Athens Charter, numerous social and cultural facilities.

The non-standardized architecture of its 900 different dwellings (including a few detached houses and duplex apartments) was designed to be completed once the residents (about 2,270 at present) had taken ownership by greening their dwellings. Garden terraces aim to "promote the socialization of the residents, insofar as they are visually interconnected" ([25]: 54) and insofar as exchanges around the plantations are catalysts for social exchanges (Fig. 7.3).

In June 2014, an eco-citizen's charter, carried by a group of residents, defined the gardens as a "common good", a "landscape and architectural heritage" and called for the resilience of biodiversity but above all for an inclusive neighbourhood life.

> *The translated eco-citizen charter for the Maladrerie proposed on 21 June 2014 by*
> *the association Jardins à tous les étages (*Gardens on all floors*),*
> *the Confédération nationale du Logement (*National Housing Confederation*),*
> *the Collective of neighbourhood residents and the neighbourhood coordination.*
>
> **Commitment n°1**
>
> We undertake to respect the exceptional character of the Maladrerie: only discard garbage in dedicated spaces, do not disfigure our facades (tarpaulins,

Fig. 7.3 Examples of buildings with garden terraces **a** and green spaces **b** in the Maladrerie

satellite dishes, laundry on railings), do not change the original urban functioning of the neighborhood (no cars, no motorcycles or scooters), etc.

The landscape and architectural heritage of the Maladrerie is a common good, we are committed to preserving it under our collective and individual responsibility.

Commitment n°2

We are committed to preserving biodiversity in our neighbourhood. We respect the flora: no trampling, no wild cuts, no gathering. We respect the fauna: no mistreatment. Biodiversity is a vital need, and we are committed to fostering it. Nature is invaluable, we are committed to respecting, sharing and promoting it.

Commitment n°3

Our garden terraces are vital and indispensable, we are committed to maintaining and promoting them. We are entitled to beauty…
… Let us preserve the quality of life that our gardens provide.

Commitment n° 4

Let us show solidarity and let our neighbours and visitors enjoy the quality of the landscape that La Maladrerie offers us. Let us defend it!
Let us respect the view of our neighbors, keep our gardens clean.

7.2.3 Neighbourhoods Resilient to the Metropolitan Crisis of Working-Class Areas

The garden suburb of Stains experienced a first crisis from the 1950s, when the construction of large projects was favoured, nationally and internationally, over the concept of garden cities and due to a lack of maintenance of housing, which was neglected and not up to the standards of modern comfort.

Today, the crisis that these two historically working-class neighborhoods face as their inhabitants seek to be resilient is part of the new stage of capitalism. It is the metropolitan crisis faced by many such working-class neighborhoods. In both neighbourhoods, unemployment and social precariousness run high (over 20% of the working population is unemployed, especially young people; in the Maladrerie, 40% of the population live under the poverty threshold).

In addition, national housing policies tending to reserve social housing for the poorest, take their toll on these neighbourhoods, which experience daily problems. Before its renovation, the garden suburb of Stains saw illicit uses of communal gardens and intrusions into private gardens. In the Maladrerie, some gardens are abandoned or even de-vegetated, and trafficking has been rife since the 1980s, but the neighborhood has since become quieter.

However, these experimental working-class neighborhoods have met this metropolitan challenge with a notable resilience, thanks to the recent heritagization.

7.3 The Recent Heritagization of Two Social Housing Neighbourhoods

7.3.1 In the Garden Suburb of Stains, an Heritagisation by Appropriation, by Designation [26] and by Local Authorities

In the garden suburb of Stains, the heritage process began in 1976 with institutional protection, the inclusion in the inventory of the picturesque sites of the Seine Saint Denis. Mostly, however, the resilience process was initiated from the beginning of the 2000s throught an heritagisation by appropriation: the tenants Association, very active at the time, petitioned the Heritage service of the Seine-Saint-Denis department Council and the landlord for an exhibition and rehabilitation.

This led Plaine Commune to make this suburb a jewel of its tourism and heritage policy. In 2008, as part of its tourism plan, it opened an interpretation room *Mémoires de cité-jardin (Memories of a garden suburb)* in an old hardware store, which became

Fig. 7.4 The interpretation room *Mémoires de cité-jardin* (*Memories of garden suburb*) in Stains

a structuring place for local social life and has since welcomed some 9,000 visitors, about half of them inhabitants of Stains (Fig. 7.4).[6]

This experience is recognized by heritage institutions on several scales: in 2014, this suburban intercommunality typical for French "banlieue" was labeled *Ville d'Art et d'histoire* (City of Art and History), the local *Memories of garden-suburb* thus becoming one of the poles of the Center of interpretation of architecture and heritage that Plaine Commune is currently creating in network[7]; the garden suburb won the European Union Cultural Heritage Prize in 2015 and of the Label *Patrimoine régional d'Ile de France* (Ile de France Heritage Label of regional interest) in 2018.

7.3.2 In the Maladrerie, a Deep Heritagization by Appropriation of Residents

The Maladrerie is the subject of a strong heritagization by appropriation of its residents. In 1995, architects who participated alongside R. Gailhoustet in its design, who still practice and live in the large housing complex, created the association *Jardins à tous les étages* to ensure the resilience of the maintenance of garden terraces (Fig. 7.5).

It currently counts about fifty of the original residents and about twenty more recently arrived, plus a dozen exterior members, but influences a much larger part of the inhabitants, in that it acts to support the neighbourhood. The association campaigns against the paving of terraces, some of which suffer from waterproofing

---

[6] Source: city of Stains, http://www.stains.fr/divertir/culture/cite-jardin/.

[7] In 2018, a prefiguration study was conducted.

Fig. 7.5 Headquarter of the association *Jardins à tous les étages* (Gardens on all floors) in the Maladrerie

problems, and for inclusion in the additional inventory of historic monuments of the area.

A few years earlier, in 2008, this large atypical social housing estate was recognised under the label *Patrimoine Xxe siècle* (Heritage of the twentieth century),[8] created in 1999 by the French Ministry of Culture and awarded by the Regional Directorate of Cultural Affairs of Ile de France[9] [27]; it does not however provide legal protection. File of the central part of the Maladrerie is at the end of June 2023 about to be submitted to obtain the registration in the additional inventory of Historical Monuments at the initiative of a collective of associations and of inhabitants created in 2021 with the agreement of the municipality. Perhaps in 2024 or 2025 a part of this original great housing estate will become a protected heritage.

---

[8] Transformed in July 2016 into the label *Architecture contemporaine remarquable* (Outstanding Contemporary Architecture).

[9] Which then labeled, not without initial reluctance, 40 housing projects from the period 1945–1975, on the model of the study conducted by the Regional Directorate of Cultural Affairs of Provence-Alpes-Côte d'Azur on « housing estates and residences in Marseille 1955–1975».

7.4 A Model of Popular Urban Resilience Based on Heritagization

These two neighbourhoods present a model of popular urban resilience based on heritagization.

7.4.1 The Resilience of a Quality Urban Environment, Both Built and Green, for Residents

This resilience is first the resilience of a quality urban built and green environment for all.

Between 2006 and 2020, the garden suburb of Stains underwent a "rehabilitation respectful of the heritage and the inhabitants"[10] carried out by its landlord *Seine-Saint-Denis Habitat,* under the control of the *Architecte des Bâtiments de France*[11] (Architect of French Buildings) and with the active participation of the residents (especially mothers, with an association called the "Committee of Moms" who campaign for the implementation of children's games and sports equipment) and the intercommunal authority.

It has improved the comfort and energy performance of housing, refurbished roads and interior courtyards (Figs. 7.6 and 7.7).

In the Maladrerie, rehabilitation is still in its infancy. The first National Urban Renewal Program (ANRU[12] 1 from 2004 to 2020), was influenced by the association *Jardins à tous les étages* and limited itself to rehabilitating the Daquin islet. As part of the current National Urban Renewal Programme (ANRU 2 since 2014), the 2018 development plan was preceded by a heritage diagnosis in 2016. In 2017, the association *Jardin à tous les étages* obtained, in an unprecedented way, to become a stakeholder: a protocol signed with the National Agency for Urban Renewal, the actors of housing and the local community aims to ensure the resilience of the terraces-gardens by avoiding their mineralization. In spring 2021, the Architect of the Buildings of France of Seine-Saint-Denis guaranteed that the heritage value of the site, identified and visited by the Docomomo Association,[13] would be taken into account in its rehabilitation by 2028.

---

[10] To translate in english the title of a conference held in 2019 at the garden suburb of the Butte Rouge in Châtenay-Malabry (https://www.citesjardins-idf.fr/events/conference-la-cite-jardin-de-stains-une-rehabilitation-respectueuse-des-habitants-et-des-batiments/).

[11] He is the representative of the State at departmental level that controls the evolution of the heritage districts.

[12] *Agence Nationale pour la Rénovation urbaine* (National Agency for Urban Renewal).

[13] DOCOMOMO International is a non-profit organization dedicated to documentation and conservation of buildings, sites and neighborhoods of the Modern Movement initiated in 1988 (source: DOCOMOMO). The French working group was created in 1991.

7 The Heritagisation of Social Housing, a Model of Ordinary Urban … 117

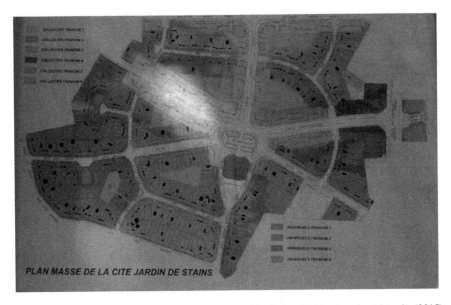

Fig. 7.6 Map of the rehabilitation of the garden suburb in the local *Mémoires de cité-jardin* (2015)

The architectural and urban resilience is however partially threatened today: the new right-wing municipality of Aubervilliers has modified the urban renewal project by reducing its budget, plans some demolitions, an opening in the city and its conversion into residential units. Since the fall of 2021, this new project, which has been developed without any consultation, has generated a lot of opposition locally, in which *Jardins à tous les étages* actively partakes. But the strong mobilization of inhabitants has already achieved many successes and the authorities become more and more conscient of the heritage value of this architecture.

7.4.2 A Symbolic Resilience of Social Housing Neighbourhoods

Resilience through heritagization also includes a symbolic dimension.

At the beginning of the development of the garden suburb of Stains, some inhabitants express doubts: "We are not in an Indian reservation, what do you come to see? Is it the poor you come to see?".[14]

Since then, regular guided tours, attracting thousands of visitors, mainly from Seine-Saint-Denis and the Ile de France and some national and international tourists, as well as artistic actions, with the active and frequent participation of residents [28],

---

[14] Source: interview with the head of tourism promotion of the garden suburb, April 2013.

(a) with games for children

(b) with shared gardens

Fig. 7.7 Two interior courtyards redeveloped by Plaine Commune and the Public Housing Office at the instigation of the inhabitants in public spaces with games for children **a** or shared gardens **b** (2015)

aim to nuance or even reverse the French depreciative imaginary of the "banlieue" and of social housing, with a certain local and media success.

As early as 2013, the tourism officer explained to the residents "the value of the place in which they live, while not hiding the difficulties of everyday life". A heritage and tourist signage has been installed, disseminating the presentation of urban space like an open-air museum outside the historic centers and hybridizing daily life with the tourist (Fig. 7.8).

From its conception, the Maladrerie has included some fifty artists' workshops, many of which are involved in the redevelopment of the neighbourhood through exhibitions, shows, installations and testimonies (Fig. 7.9).

Resilience thus depends on heritagization and "artialisation" of these social housing units. Heritagization also contributes to the transmission of achieved utopias of social housing, which combine a social project with quality architecture.

Fig. 7.8 Heritage and tourist signs and guided tour of the garden suburb of Stains

Fig. 7.9 Landscaper Sylvie da Costa's orchard in the Maladrerie (2015)

7.4.3 The Resilience of a Social Model

So the key lies in the resilience of a social model: that of working-class neighbourhoods with many social and cultural facilities (including the Paul Eluard space and theatre on the central square of the garden suburb of Stains, the Renaudie space and the Aubervilliers Center of Plastic Arts, exhibition organizer in the Maladrerie) and actions (including during the COVID-19 pandemic lockdown), in favour of the residents and especially the schoolchildren.

The garden suburb of Stains is a success, a «working-class but not insecure» neighbourhood according to the official in charge of tourism, to which the inhabitants are attached. It has become attractive again after rehabilitation while maintaining affordable rents and constituting the municipal centrality. The renovation of the neighbourhood also put an end to most of the intrusions, degradations and illicit uses it previously suffered.

In the Maladrerie, the association *Jardins à tous les étages* claims a "right to beauty" for all. While acknowledging in its Bulletin that it cannot "fight the injustices, unemployment and precariousness that afflict the population", it seeks to give back to inhabitants, some of whom are very attached to the neighbourhood, "dignity, a desire to live together, a sense of commonality". In line with the booklets distributed to the original residents, gardening or DIY workshops or plant swaps are regularly organised, so that the residents can better inhabit this atypical social housing complexwhich, surrounded by large, standardised social housing blocks, forms « the

16th arrondissement[15] of Aubervilliers»,[16] and strengthen the local sociability and the social bond.

It invites residents to cultivate their garden, literally and philosophically, thanks to the educational space *Terrasses d'avenir* ("Terraces of the future"), opened in 2016 in an apartment provided by Aubervilliers housing Office. **Thus, heritagization works together to ensure the resilience of buildings, of revegetation, of a way of life and of the ideal of a "social role of architecture".**[17]

7.5 Conclusion: Popular Urban Heritagization and Resilience

These two examples, **emblematic of the "matrimonialization"**[18] **and the "global garden"**[19] **trends that manifest themselves in the contemporary paradigm of "vulnerable omni-heritagization"** [29] show that heritagization can ensure popular urban resilience.

In some cases, heritagization of working-class neighbourhoods has been instrumentalized in favour of a "heritaglobalisation" [30] and of a metropolitan bifurcation conducive to gentrification, as in New York in the Lower East Side following the aperture of the Tenements Museum [31]. This museum inspires the current project of the *Association pour un Musée du Logement populaire* (AMULOP[20]: Association for a Museum of popular housing) created in 2014, which brings together researchers in social history and/or history teachers working in the territory of Plaine Commune in search of a nineteenth century housing building to establish such a museum, but anxious to avoid any instrumentalization or further gentrification.

A prefiguration exhibit, "Ma vie HLM" (My life in social housing), takes place between October 2021 and June 2022 in a housing block of the city Emile Dubois

---

[15] Parisian district symbolic of the wealthy upper class.

[16] Source: participant observation in a guided tour of the exhibition "La vie HLM" (The life in social housing) in March 2022.

[17] Source: interview of june 2019 with an architect-inhabitant that took part in the construction of the Maladrerie, active member of *Jardins à tous les étages*.

[18] "The matrimonialization scenario focuses on the term 'matrimony', which refers to emotions and the affective dimension (…). The term also denotes a territorialized and anchored heritage (…) whilst also having a use value: it is workable, it is 'lived-in'. It is organically integrated into the life of communities. (…) In this way, we leave an elitist, institutional (…) conception of heritage and witness the development of a socially created and geographically anchored democratized heritage which integrates alterity." ([28], p. 9).

[19] "In the Global Garden Scenario, the link between heritage and the land becomes increasingly important. Heritage merges with and becomes territory. Environnemental discourses encompass vast territories, in fact, the entire planet. Heritage becomes a concern closely linked to environmental issues (…) heritage becomes a veritable vehicle, a platform for the environnemental preoccupations of society. (…). We can class heritage as landscape or as a frame for life." (Ibid., p. 10).

[20] https://www.amulop.org/. Source: interview of april 2022 with the exhibition coordinator.

(a) View on the Grosperrin block with poster of the exhibition

(b) The kitchen of the Croisille family in 1967

Fig. 7.10 The exhibition "La vie HLM" (my life in social housing) in Aubervilliers (march–april 2022)

dating from 1957, whose urban renewal is shared with that of the Maladrerie, neighboring district of Aubervilliers.

A museology approach based on ideas (and not on objects), an immersive scenography and the mediation of guides (former tenants of social housing in the Paris region) illustrate the life of 4 families of the city: an apartment is reconstituted in its 1967 state, another shows the residential trajectories of 3 families who have succeeded each other in a building. The micro-history, based on a collection of archives and testimonies of residents, aims to change the depreciative representations of the history of French working-class neighborhoods and migrations (Fig. 7.10).

The Parisian suburban experiments presented here transmit and update the urban and social model of quality social housing. They thus contribute to the resilience of the "general model of social housing" [32, 33], dominant in France between 1945 and 1973, now maintained in some localities: massive building of social housing for the greatest number, with a qualitative or even experimental concern.

These two cases are able to inspire on a bigger scale more resilient and above all differently resilient cities in metropolisation. Thus, the garden-suburbs Association of Ile de France, created in October 2015, codirected by Stains[21] and Suresnes garden-suburbs, develops a regional network, that organizes an event called the Spring of Garden Suburbs, but also a national and even international network.

Several specialists of urbanism consider that garden-suburbs in general as «an always relevant alternative of urban development» [20], «an ideal to pursue»

---

[21] The headquarter of the association is located in the local *Mémoires de cité-jardin*.

[34], whereas some inhabitants of the Maladrerie are mobilizing for a hoped-for resilience and rehabilitation, which can be observed in other districts of "combinatory" architecture.

For exemple, in the Renaudie city of Villetaneuse, partly posthumous work of Jean Renaudie also located in Plaine Commune [35], the demolition of a part of the large complex, characterized by its highly original architecture and social model, was avoided thanks to a mobilization led by the son of the architect and by the sale of the property to former tenants of social housing, appreciative of this architecture, by the company "Les Jardins Renaudie" (Renaudie's Gardens) created by the promoter DCF, close to Plaine Commune.

The municipality and the intercommunality develop pedagogical, cultural and social actions to help all inhabitants take ownership of their gardens-terraces and to better inhabit this popular heritage (Fig. 7.11).

Under what conditions can such a popular urban resilience scenario occur and potentially spread? A comparison of the two cases shows the role played by five factors in particular.

(1) **The implication of inhabitants in the heritagization and resilience of their neigbourghood.**

The sociologist Sabrina Bresson [24] puts forward a ternary typology inspired by "Renaudian" and Le Corbusier's neighborhoods: those who are "convinced" (first-time residents who appreciate the neighborhood's way of life), those who are "captives" (more recent arrivals only remaining in the neighborhood for economic reasons) and those who are "followers" (recent tenants, with average incomes, anxious to live in Renaudie housing).

Fig. 7.11 The Renaudie city of Villetaneuse

From Plaine Commune case study, the geographer Sébastien Jacquot [28] has identified the following "living figures": the inhabitants "recipients of a policy of tourism development of the heritage", the inhabitants "as witnesses", "as tourists", "as ambassadors" and/or "as co-producers".

From Toulouse case studies, the researcher in architecture Audrey Courbebaisse [36] proposes a typology of the inhabitants appropriating themselves certain great social housing complexes: "the inhabitants-guardians of the historical values of the great ensemble", the "inhabitants-activists for ecological values", "the hedonist-inhabitant cultivating the values of living well at home".

From our case studies, we could distinguish the inhabitants active in the heritagization (some inhabitants of the garden city of Stains, the members of *Gardens on all floors* in the Maladrerie), the inhabitants receptive to the heritagization actions carried out by the first group and/or by the local authorities (around half of the garden terraces are maintained in the Maladrerie) and those indifferent to heritage issues, and deep in social concerns.

(2) **Anchoring all or part of the population promotes urban resilience through heritage appropriation.**

In the garden city of Stains, a right of filiation, exceptional and paradoxical in social housing, was granted to tenants by the housing Office in the 1960s, while in the Maladrerie, some of the inhabitants have resided since construction and/or childhood and choose to stay or return. This attachment of many residents to their neighborhood makes them actors of its heritage resilience.

(3) **The intensity of neighborhood life, a true intangible heritage, also plays a determining role in people's urban resilience.**

Associative (especially in the Maladrerie[22] since its construction) and municipal social action and cultural programming are very active in both cases studied.

(4) **In some cases, local activists can co-produce popular urban resilience.**

The territory studied historically belongs to the «red suburb» of Paris. The commune of Stains is still currently ruled by the communists, but the residents of the garden suburbs met by the responsible of the tourism mission are rarely politicized. That of Aubervilliers, with the exception of a socialist mandate, has long been a communist town, as has Villetaneuse, until the last elections. The Maladrerie is left-wing stronghold, politically and for the unions. However, the mobilization in favor of the heritagization and of the resilience of this atypical ensemble goes beyond this affiliation.

(5) The **hybridization of social and heritage policies or actions**, characteristic of Plaine Commune (Djament 2020), conception of ordinary heritage which includes the inhabitants, **bears a scenario of resilience through heritagization.**

---

[22] In particular, the *Malabyrinthe* association, a "collective of initiatives made up of residents, artists and associations", also aims to "make the city shine and (to) reveal the magic of its public spaces (…) in a spirit of conviviality" (source: neighborhood blog Maladrerie-Émile Dubois, 2016) and the Association de la nouvelle génération immigrée (Association of the new immigrant generation).

Heritagization can be a transversal lever to improve the living conditions of the residents of social housing and contributes, with housing policies,[23] alternative cultural and tourism policies,[24] or with the social and solidarity economy[25] [37], to the resilience of a working-class territory.

If the potential generalization of such experiments remains to be examined on the basis of comparative studies [38], these do prove that **a popular urban resilience, residing in the way of inhabiting and keeping alive an ordinary heritage, is possible.** If there is more than one conception of what constitutes heritage, there is more than one conception of what constitutes resilience.

References

1. Halbert L (2010) L'avantage métropolitain. PUF, Paris
2. Robinson J (2006) Ordinary cities: Between modernity and development. Routledge, London
3. Sassen S (1991) The Global city: New York, London, Tokyo. Princeton University Press, Princeton
4. Ghorra-Gobin C (2007) Une ville mondiale est-elle forcément une ville globale ? Un questionnement de la géographie française. L'Information Géographique 71(2):32–42. https://www.cairn.info/revue-l-information-geographique-2007-2-page-32.htm, https://doi.org/10.3917/lig.712.0032
5. Reghezza-Zitt M, Rufat S, Djament-Tran G, Le Blanc A, Lhomme S (2012) What resilience is not: uses and abuses. Cybergeo European Journal of Geography, Environment, Nature, Paysage 621. mis en ligne le. https://doi.org/10.4000/cybergeo.25554. http://cybergeo.revues.org/25554
6. Reghezza-Zitt M (2013) Utiliser la polysémie de la résilience pour comprendre les différentes approches du risque et leur possible articulation. EchoGéo 24. https://doi.org/10.4000/echogeo.13401. http://journals.openedition.org/echogeo/13401
7. Vale LJ, Campanella TJ (dir.) (2005) The Resilient City. How modern cities recover from disaster. New York: Oxford University Press
8. Vallat C (dir.) (2009) Pérennité urbaine, ou la ville par-delà ses metamorphoses, vol 1. Traces. Paris, L'Harmattan
9. Djament-Tran G, Reghezza-Zitt M (dir.) (2012) *Résiliences urbaines. Les villes face aux catastrophes.* Le Manuscrit, Fronts pionniers, Paris
10. Paddeu F (2012) Faire face à la crise économique à Detroit : les pratiques alternatives au service d'une résilience urbaine? L'Information Géographique 76:119–139. https://doi.org/10.3917/lig.764.0119. https://www.cairn.info/revue-l-information-geographique-2012-4-page-119.htm
11. Djament G (2012) Rome, la 'Ville Eternelle' est-elle une ville résiliente? Résilience symbolique, légitimation politique, reproduction sociale. In: Djament-Tran G, Reghezza-Zitt M (eds)

---

[23] The last Intermunicipal Community Housing Program of Plaine Commune (2016–2021) provided 40% of social housing among the 4200 dwellings to be built on average per year. Source: Plaine Commune, PLH 2016-2021, action program, https://plainecommune.fr/fileadmin/user_upload/Portail_Plaine_Commune/LA_DOC/THEMATIQUES/Habitat/4-Plan_actions__PLH_2016_2021.pdf.

[24] They are partly intended for tourists in the traditional sense, for famous monuments such as the Basilica of Saint-Denis, but also, especially in the studied districts, for the inhabitants, in the narrow or the broad sense.

[25] Products derived from the garden city are sold by associations such as *Franciade*, located in Saint-Denis.

Résiliences urbaines. Les villes face aux catastrophes. Le Manuscrit, Fronts pionniers, Paris, pp 73–102
12. Gravari-Barbas M, Veschambre V (2003) Patrimoine : derrière l'idée de consensus, les enjeux d'appropriation de l'espace et des conflits. In: Melé P, Larrue C, Rosemberg M (eds) Conflits et territoires. Presses universitaires François Rabelais, Tours, pp 67–82
13. Gravari-Barbas M (dir.) (2005) Habiter le Patrimoine. Presses universitaires de Rennes, Rennes. https://books.openedition.org/pur/2208?lang=fr
14. Veschambre V (2008) Traces et mémoires urbaines. Enjeux sociaux de la patrimonialisation et de la démolition. Presses Universitaires de Rennes, Rennes
15. Mengin C (1998) Le patrimoine du pauvre: l'habitat social en France et en Allemagne. In: Andrieux J-Y (ed) Patrimoine et société. PUR, Rennes, pp 133–141
16. Veschambre V (2014) Les grands ensembles français: un patrimoine encombrant en ce début de XXIe siècle. In: Djament-Tran G, P San Marco (eds) La métropolisation de la culture et du patrimoine. Éditions Le Manuscrit, collection Fronts pionniers, Paris, pp 367–406
17. Pouvreau B (2008) Le Coin du feu à Saint-Denis (1894–1914). Une société coopérative d'HBM pionnière pour la Caisse des dépôts et consignations. Histoire Urbaine 23:41–54. https://www.cairn.info/revue-histoire-urbaine-2008-3-page-41.htm
18. Jacquot S, Fagnoni E, Gravari-Barbas M (2013) Patrimonialisation et tourisme dans la région métropolitaine parisienne. Le patrimoine, clé de métropolité touristique? In: Gravari-Barbas M, Fagnoni E (eds) Métropolisation et tourisme. Comment le tourisme redessine la métropole parisienne. Belin, Paris, pp 102–117
19. Corteville J (dir.) (2018) Association régionale des cités-jardins d'Ile de France. In: Les cités-jardins d'Île-de-France, une certaine idée du bonheur. Lieux Dits: Lyon
20. Baty Tornikian G (dir.) (2001) Cités-jardins : genèse et actualité d'une utopie. Les Cahiers de l'IPRAUS
21. Coudroy de Lille L (ed.) (2013) Henri Sellier. La cause des villes. Histoire Urbaine 37:2. https://www.cairn.info/revue-histoire-urbaine-2013-2-page-5.htm
22. Pouvreau B, Couronné M, Laborde M-F (2007) Les cités-jardins de la banlieue du nord-est parisien. Éditions Le Moniteur, Paris
23. Chaljub B (2019) Renée Gailhoustet, une poétique du logement. éditions du Patrimoine, Monum, Paris
24. Bresson S (2010) Du plan au vécu. Analyse sociologique des expérimentations de Le Corbusier et de Jean Renaudie pour l'habitat social. thèse de sociologie de l'université François Rabelais de Tours. http://www.theses.fr/2010TOUR2018
25. Niez A (2015) La figure de la terrasse-jardin dans le logement collectif. École nationale supérieure d'architecture et du paysage de Bordeaux: research dissertation directed by Ch. Bouriette
26. Rautenberg M (2003) La rupture patrimoniale. Grenoble, Bernin, À la croisée
27. Gaudard V, et al (dir.) (2010) 1945–1975, une histoire de l'habitat: 40 ensembles de logements "Patrimoine du XXe siècle. Issy les Moulineaux: Beaux-Arts Magazine
28. Jacquot S (2015) Politiques de valorisation patrimoniale et figuration des habitants en banlieue parisienne (Plaine Commune). EchoGéo 33. https://journals.openedition.org/echogeo/14317
29. Gravari-Barbas M (dir.) (2014a) Atelier de Réflexion Prospective Nouveaux défis pour le Patrimoine culturel/Foresight workshop New challenges for cultural heritage. Synthesis of the final report. Paris 1 University. https://eirest.pantheonsorbonne.fr/sites/default/files/inline-files/ARP_Synthese_anglais.pdf
30. Gravari-Barbas M (2012) Tourisme et patrimoine, le temps des synergies? In: Khaznadar C (ed) Le patrimoine oui, mais quel patrimoine? Éditions Babel, coll. Paris, Internationale de l'imaginaire, pp 375–399
31. Gravari-Barbas M (2014b) Patrimoine, culture, tourisme et transformation urbaine. Le Lower east side tenements museum, NY. In: Géraldine D-T, Marco PS (eds) La métropolisation de la culture et du patrimoine. éditions Le Manuscrit, collection Fronts pionniers, Paris, pp 141–179
32. Ghekière L (2007) Le développement du logement social dans l'Union européenne. Quand l'intérêt général rencontre l'intérêt communautaire. Dexia Éditions, Paris

33. Desjardins X (2008) Le logement social au temps du néolibéralisme. Métropoles 4. http://journals.openedition.org/metropoles/3022
34. Paquot T (2013) Les cités-jardins, une alternative d'avenir ? In: Les cités-jardins, un idéal à poursuivre. Cahiers de l'IAURIF 165, avril. 88. https://www.iau-idf.fr/fileadmin/NewEtudes/Etude_885/FR_c165_web.pdf
35. Djament G (2020) La patrimonialisation du logement social, observatoire de l'omnipatrimonialisation fragile. In: Les Cahiers de la recherche architecturale urbaine et paysagère, vol. 8. http://journals.openedition.org/craup/4776
36. Courbebaisse A (2021) Appropriations habitantes dans les espaces intermédiaires des grands ensembles toulousains. Projets de paysage 24. http://journals.openedition.org/paysage/19680
37. Cousin S, Djament-Tran G, Gravari-Barbas M, Jacquot S (2015) Contre la métropole ... tout contre. Les politiques patrimoniales et touristiques de Plaine Commune, Seine-Saint-Denis. Métropoles 17. http://journals.openedition.org/metropoles/5171
38. Auclair E, Hertzog A (dir.) (2015) Grands ensembles, cités ouvrières, logement social: patrimoines habités, patrimoines contestés/High-rise Estates, Industrial Tenements And Social Housing : Concerns And Controversies Around Inhabited Heritage Sites. EchoGéo 33. https://journals.openedition.org/echogeo/14376
39. Bresson S, Fijalkow Y, Iosa I (2020) Architecture et logement social, quels renouvellements?/Renewals for Architecture and Social Housing? Les Cahiers de la recherche architecturale urbaine et paysagère 8. http://journals.openedition.org/craup/4666
40. Coudroy de Lille L, Crespo M, Pouvreau B (dir.) (2022) Association régionale des cités-jardins d'Ile de France. In: Des cités-jardins pour le XXIe siècle. Valorisation, préservation, perspectives. Parenthèses, Architectures, Paris
41. Djament G (2019) Patrimonialisations, territorialisations et mobilisations dans la banlieue rouge: Plaine Commune et le patrimoine de banlieue. L'Espace Politique 38. http://journals.openedition.org/espacepolitique/6726

Géraldine Djament (translation in English with the kind proofreading and correction of Rosalie Delpech) Lecturer at Strasbourg University qualified to direct research in geography, Member of the laboratory Sociétés, Acteurs et Gouvernements en Europe (UMR 7363 SAGE: Societies, Actors and Governments in Europ), Associate member of the Equipe Interdisciplinaire de Recherches sur le Tourisme (EA 7337 EIREST: Interdisciplinary Tourism Research Team of the University Paris 1 Panthéon-Sorbonne) Member of the Association of Critical Heritage Studies (ACHS)

Chapter 8
The City-State of Hamburg (Federal Republic of Germany): Metropolization, Port Functions and Urban Resilience

Patricia Zander

8.1 Introduction

The radical nature of the technical, environmental, and social changes observed in the twentieth and twenty-first centuries has overturned the conventional grids of analysis of the city.

The metropolitan stage, characterized by the global networking of places and liberalism's competition, requires cities to find a new place on the world scene [1]. Cities are mutating and must renew themselves: to remain singular in an ever more complex network, to keep their specificities while adapting to this new world. To be resolutely resilient?

In this context, Hamburg, a European metropolis belonging to the Federal Republic of Germany, seems to be a relevant example to question contemporary urban resilience. This concept will be understood, in the context of this article, as responses from a system in the face of major disturbances to allow it to recover its functions [2].

Indeed, in Hamburg, actions have been carried out for more than 50 years seemingly to activate resilience processes in order to enable it to maintain its place in the world, well expressed by its multi-secular territorial motto: "*Das Tor zur Welt*" literally: "The door to the world". A European metropole for a long time uncontested, it has been weakened by successive crises. Hamburg is a city-state, an extraordinary political power in Europe, which is itself undergoing a major metamorphosis. New actors are becoming legitimate to think, invent and orchestrate space on a metropolitan scale. Development choices have been made by political actors and new tools for resilience have been tested. Among them, the *Internationale Bauausstellung* (IBA), literally "International Architecture Exhibition" is an approach promoted in

---

P. Zander (✉)
Laboratoire SAGE (Sociétés, Acteurs, Gouvernement en Europe), Université de Strasbourg, Strasbourg, France
e-mail: patricia.zander@unistra.fr

© The Author(s), under exclusive license to Springer Nature Singapore Pte Ltd. 2023
L. Zhang et al. (eds.), *The City in an Era of Cascading Risks*, City Development: Issues and Best Practices, https://doi.org/10.1007/978-981-99-2050-1_8

Germany since the beginning of the twentieth century. Originally, a vector of changes as profound as they are rapid and visible, this "instrument" allows us to question urban resilience.

After having given the following methodological approach (part 1), we shall present this city-port that has been progressively weakened since the end of the twentieth century (part 2) and then the main responses that have been made to face these challenges (part 3). Finally, we will present the IBA tool which is at the origin of new modes of intervention for new urban projects, thus questioning contemporary urban resilience (part 4).

8.2 Methodology

In the context of these investigations, the approach followed is complexity, an approach that I particularly explored in the context of the Habilitation to Direct Research [1]. Complexity in the human sciences presents different approaches, as presented for example by [3] or [4]. The approach is based on the work of Barel and his paradox approach to complexity [5] where the paradox is considered as an unsurpassable contradiction. A classic paradox–the territory/network paradox–is becoming more acute in contemporary times and is the subject of my research: "Territories are bypassed by networks and networks thicken into territories. To overcome these paradoxes, it is not useless to redefine the tools for classifying the different types of spaces" [6], p. 102).

The data mobilized comes from different sources. First of all, field observations were made in October 2015. These field observations made it possible to identify certain urban practices. They also allowed us to identify different types of territorial mutations in the central and pericentral districts of Hamburg (Hamburg-center, Altona, Pauli, HafenCity) and, to the south of the Elbe, on the Island of Wilhelmsburg. During this excursion, the information centers for the public and researchers in HafenCity and the IBA in Wilhelmsburg were used to examine technical aids (models, maps). Finally, the Hamburg metropolis is well documented scientifically and technically. Access to statistics is facilitated by websites such as https://www.statistik-nord.de or https://metropolregion.hamburg.de/.

Similarly, blogs opposing the metropolitan project are rich resources: they multiplied between 2007 and 2013, during the course of the Hamburg IBA, and make it possible to grasp the limits of the actions carried out, the criticisms and expectations of a part of the population confronted with the metropolitan changes.

These tools and sources were mobilized in cooperation with a Strasbourg architect-urban planner, a specialist in local development (Guy Giraud). The interest of this complex approach is the unprecedented crossing of approaches and data, allowing a fresh look at spatial realities, particularly urban ones. It is thus possible to concretely implement a multidisciplinary approach to urban issues.

8.3 Hamburg, A Weakened Port Power

Hamburg is intriguing. The contradictions of metropolitan processes are bluntly expressed: Hamburg's economic indicators show the insolent success of capitalism but also its corollary, the exacerbation of social polarities in a German context dominated by the rise of extreme poverty.[1] Through its port and long-distance trade, Hamburg concretely embodies the city's "network dimension". The port feeds the metropolis, despite its weakening: its share of the economy is statistically declining, yet it remains omnipresent, displaying its name and its "values" in new metropolitan districts such as HafenCity, districts that are resolutely turned towards the world, as the port was and remains.

Hamburg is intriguing because network and territory are inseparable and seem more complementary than contradictory. The interweaving of attributes that make up its identity is a good indicator: it is first and foremost a port, but the port does not exist without the city, which is a state; it is a node connected to the world; the city is singular, Hanseatic, and also irreducibly German. None of these characteristics can be considered in an isolated way.

For a European city, it was born late (only in the ninth century, its close neighbor Lübeck dates from the seventh century), we will point here to a few facts and figures (all from Hamburg's statistical office, unless specifically mentioned[2]) reflecting its original position in Europe.

8.3.1 Hamburg, A World-City Born of a Port

The port is not strictly speaking maritime and the environment is not evenly favorable [7]: Hamburg belongs to the Northern Plain (Das Norddeutsche Tiefland), which extends in Europe from northern France to Siberia. Originally, the city was located about 100 km inland on an infertile terrace overlooking the Elbe (a characteristic environment of the Northern Plain, the *Geest*[3]). This terrace is not very suitable for agriculture but, thanks to its waterways, it offers an exceptional urban setting (Alster lake, Bille river). Figure 8.1 shows the spatial organization of Hamburg's different natural and urban units.

To the south of this terrace, there is a great change in environment: peat bogs are dominating the arms of the Elbe developed for the Port, forming a corridor upstream of the Elbe from Hamburg. The *Marschen* are also especially striking: these former marshes, very fertile, were drained and put into intensive cultivation since the twelfth century; they are comparable to polders. In the Hamburg area, the *Marschen* have been particularly developed for fruit production: the *Altes Land* is a huge orchard

---

[1] In 2017, 15.7% of the population was estimated to be affected, or 13 million people according to the Paritatische Wohlfahrtsverband.

[2] These figures are available at http://www.statistik-nord.de/daten/, (accessed 05/12/2017).

[3] Geest refers to low-fertility land formed from glacial deposits.

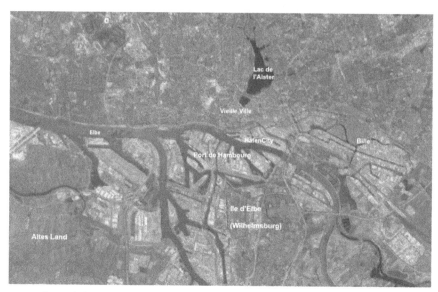

image© 2017 google, map data ©2017 geobasis-de/bkg (©2009), google france

Fig. 8.1 Landmark map of hamburg

(143 km^2, nearly a million apple and cherry trees, considered one of the largest orchards in the world), which has become an attraction, especially at the time of its spring bloom. Given the quality of its landscape and its proximity to Hamburg, the *Altes Land* is now being altered by the number of tourists and the development of its residential function. The fact that it belongs to both Lower Saxony and Hamburg complicates the institutional and political regulation of centre-periphery metropolitan relations and argues for a metropolitan political level.

The city-state has 1,899,160 inhabitants in 2019 (1,814,597 in 2012[4]), 755 km^2. Hamburg compensates its low natural growth with a steady flow of immigrants [8], thus escaping the "shrinkage" affecting some German cities [9]. Regardless of the scientific methodology used, Hamburg is unquestionably a European-level metropolis (e.g., [10, 11]).

In the ranking of European cities, Hamburg offers a combination envied by many European politicians: according to Eurostat, in 2015, its GDP per capita is the highest in Germany and the third highest in Europe (behind London and Luxembourg). Its population density is modest (2,409 inhabitants per square kilometre in 2012), compared to London (5,500 inhabitants per square kilometre) or to the exceptional case of Paris (21,000 inhabitants per square kilometre); Hamburg was also classified as a Green Capital by the European Union in 2011, a title awarded to cities for their environmental quality and the political actions carried out in this field.

---

[4] https://www.demografie-portal.de/.

However, as a result of metropolitan processes, the wealth gap is constantly widening. Poverty affects a significant fraction of Hamburg's population, and the unemployment rate, the portion of people receiving social benefits, and the population without professional qualifications and training are higher, on average, than in the rest of Germany [12].

8.3.2 A Complex Social and Political Structure: The Power of Political Power and Its Contestation

The Federal Republic of Germany has three city-states: Bremen, Berlin and Hamburg. The political power of the latter is an urban fact, in the full sense of the word: Hamburg prides itself on being one of the founding members of the Hanseatic League, which is at the origin of the prosperity of its port; an imperial free city in 1515, its continuous urban growth in the nineteenth and twentieth centuries enabled it to absorb the peripheral municipalities. In 1937, it already had 1.02 million inhabitants [7]. Its status as a city-state was recognized in 1949. It had its own constitution, the first chapter of which, "Die staatlichen Grundlagen" ("The State Foundations"), gave it its name and symbolically established its independence: "Free and Hanseatic City of Hamburg" (Freie und Hansestadt Hamburg). With this constitution, approved in 1952 and revised in 2012,[5] Hamburg became a state that controlled the urban area, had its own autonomy to conclude international agreements, and guaranteed equal rights (in particular parity between men and women, which appears in the fundamental articles). Its various national and international bodies (Parliament, Consulate,[6] ministries, municipality, etc.) constitute an extraordinary political and administrative density for a territory of only 755 km².

But political power cannot be reduced to institutions alone, even if democratic. Let us remember that German society is remarkable for the importance of its associative movements (Hanseatic League, trade unions, civil associations, etc.), demonstrating this individual and collective capacity for self-organization and action. Furthermore, in Germany, the associations have acquired a monopoly in the provision of territorial public services [13]. The intersection of a culture of association and opposition leads to a social dynamic invested in political changes, as seen by the communist upheavals in Hamburg in the twentieth century [14] or the anti-authoritarian educational experiments during the Weimar Republic [15] [1936]). Hamburg is indeed traversed by the "culture of protest" which corresponds "to a social practice inscribed in the long

---

[5] It would be interesting to analyze the changes that took place in 2012 and their articulation with the reforms of the Federation that were carried out in 2007. Hamburg's constitution is online: http://www.hamburg.de/contentblob/1604280/5e354265cb3c0e3422f30f9184608d9d/data/verfassung-der-freien-und-hansestadt-hamburg-stand-2012.pdf (accessed 10/12/2017).

[6] Hamburg is the second non-capital consular city in the world, after New York. Cf. Embassy of France in Germany, 2012.

term" ([16], p.128). This cultural contestation implies a conception of society integrating plurality of visions and, above all, the existence of an opposition. For example, since 1971, the Hamburg parliamentary political system has been constructed in an original way to guarantee the existence of the opposition within it ([8], p.316); it would be necessary to deepen this in order to better understand the Hamburg society and its political culture. Individuals, groups, institutions, in order to contest or take into account contestation, must first accept the principle of a society functioning with oppositions and be able to think of it as such.

This principle of opposition, and one of its social forms–contestation–are clearly manifested in Hamburg during conflicts concerning the appropriation of space and its financialization. The "right to the city", as formalized by Lefebvre [17], has a particularly strong echo here [18]. Since the beginning of the 2000s, the confrontation has become very harsh, expressing a radicalization of the opposing parties: we can cite, among others, the demonstration for the maintenance of the self-managed Rote Flora theater and its 620 injured in 2013 [16], or, more peacefully, the resistance of the inhabitants of the Esso-häuser[7] real estate complex. Indeed, in Hamburg, resistance to "deterritorializing" urban projects (Theater Rote Flora is for example an active site of this resistance) creates collectives of individuals united by a community of thought. This resistance takes various forms, reaches a very high degree of coordination and self-organization: occupation of premises or neighborhoods, artistic demonstration with an international echo (for example in the context of the G20 summit in Hamburg in July 2017, with the street performance "*1000GESTALTEN*"[8]). These movements go far beyond contesting a system that is becoming ever more unequal. It is about a periphery seeking other forms of existence, which are based on different values and norms than those provided by the center and embodied in part by political powers, private actors and the Hamburg city center.

---

[7] Esso-hauser after World War II along the Reeperbahn axis, the "historic" axis of the Sankt-Pauli neighborhood, known for its nightlife. Promised to be demolished to make way for luxury buildings, it has been the object of an organized, unexpected resistance from the tenants. These inhabitants are far from the clichés that could suggest the district: generally of modest origin, they are deeply attached to the atypical life of their district. A well-documented film preserves part of the memory of this protest "buy buy st. pauli, über die kampfe um die esso-hauser" (the title of the film is worded without capital letters) by Irene Bude (2014).

[8] 1,000 artists walked the streets of Hamburg, silent, atonic and with clothes covered with a clay coating similar to ash, symbolizing the depersonalization and death of the human being imposed by the capitalist system. The show ended with the awakening of these "living dead" (screams, undressing...). The repercussion of this demonstration was of international dimension thanks to the media coverage of the G20 summit (Cf. for example, the article in the Washington Post: online https://www.washingtonpost.com/video/world/g-20-protest-brings-1000-zombies-onto-streets-of-hamburg/2017/07/05/688eeb00-619b-11e7-80a2-8c226031ac3f_video.html?utm_term=.3c21d9b46362. (accessed 10/10/2017).

8.3.3 The Port in the Twenty-First Century: A Weakened But Decisive Function for the Functioning of the City-State

The economic structure of Hamburg is dominated by services (including the merchant sector, finance, transport, etc.), with the production sector representing only 18% [19]. The Port seems to be in retreat from an economic point of view, particularly in terms of employment: the sectors in development are now education, design and medical [20].

However, the Port alone still represents 40,000 jobs and its surface area is 87 km^2 [21]. Its draught has become its Achilles' heel, not allowing it to accommodate giant container ships, hence the development of the port of Wilhelmshaven some 100 km away, subsidized by the Federal State [22]. It remains an essential public player, representative of the intertwining of the economy and politics: the Port of Hamburg is under the jurisdiction of the city, and therefore of the Land, it has a legal personality and an autonomous budget. It is considered to be a private company but is financed by public capital.

Intermodality is its strength, favouring a solid European hinterland: connection to the German motorway network, rail with 160 international container trains departing and arriving per day ([21], p.25), connection to the European river network via the Elbe: 870 km of navigable waterways accessible to serve Germany, the Czech Republic and Poland; finally, feedering (transshipment by low-tonnage ship) is developing links with the Baltic countries and the British Isles.

Port activity is nonetheless fragile, subject to global crises (2008 in particular) and to competition: with Asian shipowners[9] and the port of Wilhelmshaven, container traffic is struggling to return to its pre-2008 level. In 2015, Hamburg lost its 2nd place to Antwerp,[10] despite ongoing developments and modernization.

Logistics, aeronautics and the media sectors are the pillars of the Hamburg economy, with their flagship European German companies: Airbus, Olympus, etc. Clusters have been developed to strengthen these foundations: *Hamburg Aviation* (40,000 employees), *Logistics Initiative Hamburg* (500 members, a cluster that seeks to strengthen the position of the port of Hamburg in the world), *NextMedia Hamburg* (created in 1997). Several clusters have been created since 2008, either to promote the development of promising economic activities (health, environment), or to strengthen the regional role of the Hamburg cluster (Maritim Cluster Northern Germany). For example, Life Science Nord brought together 12 universities and 500 companies in 2012. This cluster policy is actively supported by the City and the State Ministry of Economics, which has set up a "cluster of clusters" to promote interaction between these economic organizations.

---

[9] Cf. the Port of Hamburg website (https://www.hafen-hamburg.de) and the daily Le Monde, 10/02/2017, C. Boutelet. "In the port of Hamburg, seafarers who are disappointed".

[10] Cf. Port of Antwerp, Facts and Figures, 2016. Online: http://www.portofantwerp.com/sites/portofantwerp/files/20160714_POA-1833_Cijferboekje2016_11824_FR.pdf (accessed 05/12/2017).

Although its direct contribution to the economy is decreasing, the port continues to play a central role: it seems to be transfiguring into new international activities. Since the late Middle Ages, thanks to the Hanseatic League, Hamburg has played a direct role in the various phases of globalization, enabling trade between Europe and the rest of the world. Its historic warehouse district (*Speicherstadt*) and business district (*Kontorhaus*) dating from the late 19th and early twentieth centuries have been designated a UNESCO World Heritage Site. With *Kontorhaus*, Hamburg boasts Europe's first business district, with 300,000 m^2 of floor space dedicated to transactions.[11] Hamburg's society and its space, which are permeated by the spirit of capitalism, are adapting to the new economic and social conditions of the world, and tools are being put in place to enable this evolution.

8.4 From the City-Port to the Metropolis: A New Local and Regional Anchorage to Recover Weakened Functions

8.4.1 The Twentieth Century, a Century of Disasters for Hamburg

In view of the disasters that have marked its recent urban history, Hamburg has repeatedly demonstrated its capacity for resilience: like many European cities, it was destroyed by a major fire (in 1842, 25% of the city was destroyed: [7], Operation Gomorrah, carried out by the Allies in 1943, targeted the port, which had become a centerpiece of the Nazi military system: the city was the most destroyed in Germany and a generation of Hamburgers was traumatized by an unprecedented climatic phenomenon created by the deluge of bombs: hurricanes of fire devastated the harbor, killing 45,000 people and leaving more than a million homeless [23], deadly floods in 1962 leave 60,000 people homeless [24]. And yet, Hamburg recovered each time, and its port was rebuilt. In the 1980s, when the crisis was affecting the Hamburg economy and relegating it to a middle position in Europe, a city-enterprise policy favored this diversification of clusters that are participating in the renewal of the Hamburg economy. The future "is no longer at sea but on land" ([25], p. 292). In 1983, "Enterprise Hamburg" (Unternehmen Hamburg) marked a reorientation of the city's policy with the aim of strengthening the tertiary sector and improving the city's attractiveness and image" ([18], p.9).

---

[11] Cf. ICOMOS technical evaluation mission n°1467, carried out in the context of the UNESCO World Heritage listing.

8.4.2 Going Beyond Local Borders in the Twenty-First Century: Local Political Redeployment as a New Basis for a Metropolis on a European and Global Scale

For Hamburg, continuing to be part of the new global networks of the twenty-first century is as much a challenge as it is a work practiced several times throughout its history.

This inclusion in the networks requires a transformation of its economy, which imposes a political and urban redeployment.

Creating a metropolitan political level is an obstinate construction that began at least 60 years ago. As early as 1955, Hamburg, as a state, concluded agreements (Land Schleswig–Holstein and Lower Saxony) with its neighbors to coordinate shared peripheries [26]. The system set up at the time favoured the decision-making powers of the *Länder, Regierungsbezirke and Kreise*, to the detriment of the municipalities. This cooperation became more extensive in the 1990s with the intensification of political relations between three *Länder*, Lower Saxony, Mecklenburg-Vorpommern and Schleswig–Holstein, and the creation of the *Metropolregion Hamburg* (1995). It is a structure without legal capacity. It is the result of an international treaty between the different *Länder*, which defines the rules of financing (mobilized funds), and of cooperation agreements between the Kreise and, in part, the Communes, which determine the objectives, the working methods, and the distribution of financing. This process implies a profound renewal of the political actors, their place and their relations: some 1,000 municipalities of the Hamburg metropolis are now part of this metropolitan web, as is the private sector, represented in the Councils of this structure. A large number of texts (agreements, programs, etc., listed and easily accessible in *Metropolregion.de*) are drawing new relationships between actors, new policies with their new ground realities. Figure 8.2 below shows the central location of Hamburg, the metropolitan hub of a metropolitan region with more than 5 million inhabitants.

With this metropolitan level, Hamburg is positioning itself as a metropolis open to the Baltic. This renews its traditional role as a Hanseatic city. But its impact on the spatial organization of the neighboring Länder is contrasted, implying both an economic dynamization for polarized (and peri-urbanized) sectors and an impoverishment of other sectors [27].

An interesting slogan defines the strategic project of this structure: "Metropole Hamburg–Wachsende Stadt" which can be translated as "Hamburg, a booming metropolis" ([28], p.13). It seems more accurate to translate this slogan as "Metropolis Hamburg–Booming City", which sums up the German, and particularly the Hamburg, urban spirit quite well: taking advantage of metropolization to allow for an intensification of the city, wherever it is in the metropolitan space. We shall see that this idea is repeated in the city-state's urban project.

An efficient metropolitan space for everyday life: in Hamburg, F. Ascher's metapole seems to be an essential condition for establishing the metropolis internationally. This metapole should offer fair living conditions for individuals and

Fig. 8.2 A "city" at the center of a metropolitan area of 26,000 km^2 and 5.1 million inhabitants (2014). *Source* 2018: https://metropolregion.hamburg.de/karte/

communities; it should also enable metropolitan economic efficiency and the integration of companies, employees, and citizens in the Hamburg metropolitan region. Cooperation between research centers and universities is being promoted in order to develop knowledge of the area and its problems. One example is a research program conducted by the Technische Universität Hamburg-Harburg concerning mobility and access to services within the metropolitan region and the inequalities that may exist.[12] The objectives, the action programs 2017–2020 and their financing should be explored to be confronted with metropolitan processes such as land consumption for residential or economic purposes.

---

[12] On this subject, one can consult: http://metropolregion.hamburg.de/mobilitaet/4401134/lp-erreic hbarkeitsanalysen/ (consulted on 10/12/2017) presenting the project and an important cartography of this issue.

8.4.3 Hamburg's Urban Intensification as the Driving Force of the Metropolregion Hamburg: Spatial Development and Symbolic Exacerbation of the Center

The state of Hamburg occupies a special place in the metropolitan region: it represents the geographic and symbolic heart of the metropolitan region. The question of land is essential: how can the center retain its identity and attractiveness? How to take advantage of (and profit from) this central land?

During the twentieth century, the city grew towards the north. The development plan for 2030 foresees a brake on this expansion, recommending.

- An extension of the hyper-center with the new HafenCity district. The hyper-center continues to grow in commercial density, with a large network of shopping malls and shopping centers that blend into the urban fabric.)
- Urban intensification of part of the pericentral districts.
- The urbanization of the Island of Elba: "the jump over the Elba".
- A green belt and green corridors (Fig. 8.3).

In this context, the new HafenCity district will play an exceptional role. It is a contemporary incarnation of the port, of this port function which has been the identity of the city since the eleventh century. Remarkably well located in the heart of the city, this strategic space is now the interface between the North of Hamburg (the preferred place for urban development, and in particular its old center), the current port and the expanding south.

These 157 ha, allowing an increase of 40% of the surface area of Hamburg's city center, were at one time intended for total demolition. Their development was approved in 2000 by the Hamburg Senate on the basis of a general urban development plan.

Figure 8.4 below shows the location of the area in the southeast of the old town.

The land was used intensively, as shown by this evolution of the projected and then realized gross floor area: *"The plan was redesigned to gain space: the land area increased from 123 to 127 ha. The gross floor area of 1.5 million m2 has increased to 2.32 million m2 of gross floor area."*[13] The budget is 7 billion, two thirds of which is provided by private funds. Table 8.1 provides additional figures.

This district stands out as:

- A place of heritage and identity with the Unesco World Heritage listing of the "Speicherstadt" warehouses (370,000 m2, built between 1885 and 1927) and what is considered by Unesco as the first European business district: Konterhaus.
- A place where technical innovations were tested in terms of development, in order to preserve the district's link with the water (8 to 9 m above the sea).
- A new place for its governance.
- A place of knowledge production, with the establishment of new premises for a university and schools of all levels.

---

[13] Cf. http://www.HafenCity.com (accessed on 15/12/2015).

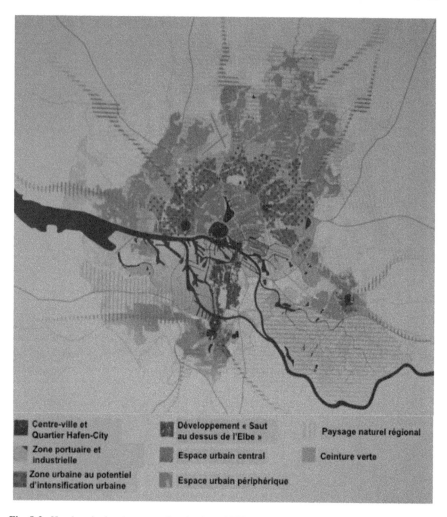

Fig. 8.3 Hamburg's development plan–horizon 2030. *Source* Hamburg, urban development and environment department

- A place that, with its offices, high-end housing and leisure activities, is intended for the development of very high-level economic activities, focused on innovation.

HafenCity is an intertwining of public spaces and publicly accessible private spaces, an expression of the partial privatization of urban centers and pericentres, as the map above clearly shows. The interweaving of public and private can be seen in this district, as in the renovation of Hamburg's peri-urban districts: for example, in the western district of Altona, the start of the rehabilitation of this area was given by the establishment of an IKEA store: it is the first of its kind by this company, whose sales outlets are mostly located in the suburbs. The IKEA company took charge of

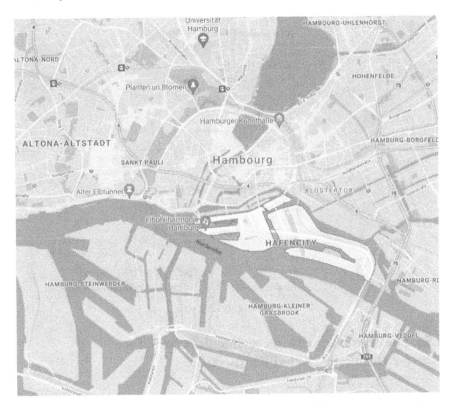

Fig. 8.4 The Hafencity district and its strategic location at the interface of the historic center, the twentieth century city and the port. Maps data ©2021 Geobasis-DE/BKG (©2009) Google

the renovation of the important public square, enhancing the value of the real estate in the vicinity and encouraging renovations or reconstruction [30]. Such operations induce a transformation of the neighborhood's settlement and deep social and urban mutations. As demonstrated by Minton [31], the contemporary urban public space is the result of a democratic battle fought in the nineteenth century. Today, this space is regressing, becoming private space accessible to the public (the Docklands being the first large-scale operation carried out in Europe).

The Elbphilbharmonie, a former cocoa warehouse, has been redeveloped at great expense in HafenCity. It has become an exceptional private/public space for Hamburg: it is a large concert hall, but also upscale private housing,[14] a luxury hotel, and parking lots, symbolic of the social and spatial changes at work. This district expresses the renewal of values and symbols that have determined the development of the city-state, for a long time incarnated by the port.

---

[14] Some of these have been put up for sale at over 18,000 per m2: https://www.engelvoelkers.com/de/ (accessed 12/12/2015).

Table 8.1 Hafencity: figures

Total surface housing, 157 ha of former port and industrial area	
Distance from the center of HafenCity to the City Hall	800 m
Extension of the area of the city of Hamburg	40%
Gross floor area of new buildings	2,32 millions de m2
Building density (land use factor)	3,7 à 5,6
Average population density	95/ha
Average working population density	357/ha
Employment	45 000 postes de travail (35 000 dans des bureaux)
Housing	6 000
Inhabitants	12 000 environ

Source HafenCity [29], p.67

8.5 The Hamburg IBA Experiment: A New Tool for Resilience? What Prospects?

8.5.1 Why an IBA in Hamburg?

The Internationale Bauausstellung, literally "International Architecture Exhibition", has been promoted in Germany since the early twentieth century. Hamburg used the IBA tool from 2006 to 2013, in parallel with the development of HafenCity. These initiatives were taken in the early 2000s, at the same time as the status of metropolis, decided at the federal level in 1995, was being asserted. From then on, the city-state of Hamburg had to consolidate its role as a regional center in a metropolitan environment that was growing in strength institutionally. A city like Schwerin, on the periphery of the metropolitan region to the east of Hamburg, recognizes the importance of the metropolitan fact and the role that Hamburg must assume: "A metropolitan region is the engine of social, cultural and technological development. Hamburg in particular, with its universities and leading companies in the fields of IT, media and design, is an innovation engine for the entire metropolitan region."[15] The scarcity of land is a major issue for Hamburg, which wants to retain its role as the heart of the metropolis. The city-state suffers from the small size of its territory and the constraints of its natural environment, which is marked by water (tides, flooding). The IBA allows the partial

---

[15] Cf. the website of the City of Schwerin, presenting the metropolitan structure, Metropolregion Hamburg, https://www.ihkzuschwerin.de/standortpolitik/Regionale_Kooperation/Metropolregion/Ueber_die_Metropopolregion_Hamburg/3031772 (accessed 5/12/2017).

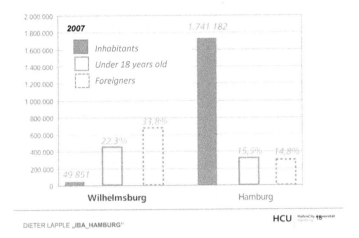

Fig. 8.5 Social structures comparison willhelmsburg-hamburg

conquest of the island of Wilhelmsburg, representing 52 km2, while the development has traditionally been turned towards the north. The south of the island, considered peripheral, has land reserves and an urban potential to be developed. The market garden areas must be protected, and urbanization contained. Hence the name given to this IBA: "Sprung über die Elbe".

The abandoned spaces of Wilhelmsburg had become in the twentieth century margins welcoming what is undesirable: landfills, major communication routes, poor, foreign populations... [32]. These spaces now offer a much sought-after potential: land to be conquered and nature that has not been (overly) tidied up and transformed by humans.

In line with this pattern, at the beginning of the twenty-first century, social relegation characterizes the island of Wilhelmsburg: about 100 nationalities are present, social dramas caught the attention of Hamburgers in 2002 [32]. The IBA has a mobilizing power and can set in motion new dynamics, capable of profoundly transforming the processes of marginalization still underway. And it can better include young populations and families, whose deficit becomes problematic at the city level (Fig. 8.5).

8.5.2 Ways and Lines of Work to Produce "More City"

The Hamburg IBA has three dimensions for stimulating new processes:

– The initiation of new, "non-routine" collective work processes, new relationships to space and between actors, including the involvement of the population.
– Concrete operations of transformation of the urban fabric while respecting the existing.

- The elaboration of background documents, allowing the transformation of the space over 10/20 years: models, plans and reports used as references.

In 2006, the company IBA GmbH was created, and three main lines of work were defined by its team of developers:

- Metrozone: opinions are divided on the concept behind this axis. Metrozone is about in-between spaces, about derelict areas, partly inspired by T. Sieverts and the Zwischenstadt. "These spaces hold the promise of a new urbanity. Density, diversity of uses, spatial potentialities, urban atmospheres" (Koch cited in [32], p.400). Metrozone refers to a category of space that, from being invisible or carrying constraints, becomes a resource for creating the city. This liberation of land has a direct effect on the environment (preservation of agricultural or natural peripheral land) and on real estate pressure, which it would help to reduce mechanically. The only condition is that the value of this land, which has become "city" land, does not contribute, a contrario, to the increase in land prices, and does not fuel the gentrification process.
- Cosmopolis (or metropolis): the increasingly urbanized space must develop its potential ("more city within the city"): here, the diversity of nationalities and cultures is a lever for promoting urban development. *Cosmopolis* is a "way" of being and a space that allows access to the diversity of the world and the development of this diversity locally.
- City and climate change: this is the environmental component of the IBA, which is unavoidable and was invited in 2007, following the publication of a UN report on climate change.

The themes and issues addressed carefully ruled out the Port, considered a "complicated" partner that could have compromised the operation ([32], p. 402). This difficulty of the relations between the actors, in a territory where the city-port osmosis is nevertheless so deep, may seem like a paradox that questions and should stimulate research.

8.5.3 Lessons and Prospects Opened Up by the Hamburg Case and the Hamburg IBA

It is hardly surprising that the IBA has served the purpose of investment and has enabled the city's capital enrichment. From this point of view, it has given rise to controversy and passionate debate. From a French perspective, there are two contradictory views:

- That of the academics (S. Depraz, C. Barbier) which is globally incriminating, confronting the project, its devices, the discourses with the realities. However, the IBA project was launched with a clear objective, which it has kept very well: to develop the island of Wilhelmsburg by attracting investors.

– The politicians and technicians were unanimous in their praise.

IBA Hamburg facts (2006–2013)	1,733 residential units built 100,000 m² of commercial space built 8 schools built 2 retirement homes 3 nurseries 4 sports centers 1 shopping center 1 center for artists and creative people More than 70 hectares of green spaces More than 420,000 visitors 700 million € of private investment and 300 million € of public funding
IBA Hamburg (2006–2013): some "intangible" operations and initiatives	The neighborhood university The knowledge center The multicultural center The educational component: education offensive (Bildungsoffensive Elbinsel) Source: https://www.construction21.org/france/articles/fr/IBA-hamburg-une-approche-allemande-du-developpement-urbain-durable.html (accessed on 10/10/2017)

Such a gap in the analyses highlights the fractures between the academic world and the political and technical world of planning.

The results, both "material" and quantitative, are undeniably flattering: the mobilization was successful, the public investment (€300,000) had, overall, the expected knock-on effect (€700,000). But above all, the IBA can (must) initiate new partnerships, be a moment of learning and renewal of a collective work that shakes up attitudes and mentalities.

For example, the social housing district Quartier Monde (Weltquartier), characterized by its hundred or so nationalities, was the subject of a consultative process over several years, the long-term effects of which should be analyzed in detail: is it simply a matter of using the resource of cosmopolitanism to enhance the district ([32], p. 403)? What about the effects of the Neighborhood University that operated from 2006 to 2013? Formed by volunteer students from the Hamburg HafenCity University (HCU[16]), this "University" occupied premises that were promised to be demolished. It was financially supported by the HCU and the IBA. "Inhabiting urban change with users allows for the development of expertise in the discipline of urban planning" ([33], p.56). What are the effects of such an initiative on the University, its knowledge production, the neighborhood, the political and technical world?

How has the IBA changed the game for the inhabitants and participated in the processes of territorialization? To what extent has it generated a structured opposition, which is at the origin of new forms of resistance and self-organization? Has

---

[16] Note that HafenCity is always written with two capital letters, valuing Port and City equally. The acronym of the University follows this writing rule.

this experimentation changed the way elected officials and technicians view the metropolis, Hamburg, and this district?

The liberal metropolis, despite its bold and innovative tools, seems to be showing its limits here: the drastic reduction in social housing, from 211,000 in 1993 to less than 100,000 in 2012 [34], is combined with a real estate market made tense by prestige developments that exclude modest households [35]. Hamburg is far from the situation in Montreal presented by Polèse [36]. Urban operations intended to enhance social housing and its environment relentlessly generate rent increases that drive out the most modest tenants (http://akuwilhelmsburg.blogsport.eu/) while SAGA, Hamburg's main social landlord, increases rents in order to consolidate the city's budget ([37], p.69).

Finally, in the race for profit that contemporary liberal society imposes, the lack of ethics of certain private actors directly generates the dilapidation of neighborhoods in Wilhelmburg: so it is with Korallusviertel, a 1970s housing estate that accommodates modest, elderly and/or Turkish populations. In 2004, the Fortress investment company took over some of the social housing in Hamburg under conditions that were supposed to benefit the tenants and the rented property. But the company did not respect its commitments and the housing stock is in disrepair, with the tenants remaining helpless in such a context.[17]

Hamburg, with the example of the island of Wilhelmsburg, unfortunately confirms a possible definition of the liberal metropolis: the greatest wealth of some rubs shoulders with the greatest poverty, another diversity than that targeted by the IBA Hamburg with Cosmospolis.

8.6 Conclusions

We can only emphasize here the omnipresence of the port, which seems to have woven the territory since its foundation, participating in its "concrescence" [38]: no resilience is possible for Hamburg without a port function, a port that is always concretely and symbolically visible. Urban resilience implies the mobilization of urban symbolics born of the concrete work of a society, its values and its organization.

A renewal of the role of politics in the activation of urban resilience raises questions: what new place does it occupy, particularly in the face of increasingly powerful private actors who know how to take advantage of the short-term rentals provided by, sometimes, centuries of social and urban constructions?

Does contemporary liberal society allow for resilience from below, thanks to its working classes, which would be included in the social fabric and would participate in feeding the social wealth of the city? Nothing is less certain, as the case of Hamburg

---

[17] AKU Blog - Arbeitskreis Umstrukturierung Wilhelmsburg: http://akuwilhelmsburg.blogsport.eu/2011-03-korallusviertel-wut-uber-politische-vernachlassigung-im-gegensatz-zu-aufwertungsbestrebungen-der-neuen-mitte/ (accessed 10/08/2021).

demonstrates, particularly because of the power of certain private companies whose only compass is profit.

But in fact, does resilience constitute a real resource for the future of cities and the planet? Or is it, on the contrary, a desolate creation of the mind [39]? According to this author, resilience forces us to focus on individuals and groups, exalting positive values (courage, solidarity, etc.) to better accept problems, catastrophes and disasters. Resilience thus appears to be a tool that avoids dealing politically with the origin of the problems. However, if we cannot contest certain political uses of resilience, it remains undoubtedly indispensable in these times of uncertainty and social and environmental disruption.

References

1. Zander P (2018) Territorialisation métropolitaine en Europe et paradoxes. Les exemples de Strasbourg et du Rhin Supérieur, vol 1. Position et projet scientifique, 492p. + annexes. Vol 2: Parcours et choix de publications, 68p. + annexes. Habilitation à Diriger des Recherches. Université de Strasbourg, Strasbourg
2. Toubin M, et al (2012) La 'Résilience urbaine': un nouveau concept opérationnel vecteur de durabilité urbaine? Développement durable et territoires. Économie, géographie, politique, droit, sociologie 3(1). En ligne. http://journals.openedition.org/developpementdurable/9208
3. Alhadeff-Jones M (2008) Three generations of complexity theories: Nuances and ambiguities. Educ Philos Theory 40(1):66–82
4. Weisbuch G, Zwirn A (2010) Qu'appelle-t-on aujourd'hui les sciences de la complexité ? Langages, réseaux, marchés, territoires. Vuibert, Paris
5. Barel Y (1989) Le paradoxe et le système: essai sur le fantastique social, 2. Grenoble: presses universitaires de grenoble
6. Lévy J (1993) A-t-on encore (vraiment) besoin du territoire? Espaces Temps Les apories du territoire 51–52:102–142
7. Reitel F (1996) L'Allemagne. Espace, économie, société. Paris, Nathan
8. Andersen U, Woyke W (2013) Handwörterbuch des politischen Systems. Springer Verlag, Berlin
9. Chatel C (2011) Une mesure du déclin démographique des villes allemandes de 1820 à 2010. Géocarrefour 86(2):81–90
10. Rozenblat C, Cicille P (2003) Les villes européennes. La documentation française, Paris
11. Taylor P, Hoyler M (2000) The spatial order of European cities under conditions of contemporary globalization. Tijdschr Econ Soc Geogr 91(2):176–189
12. Depraz S (2016b) Hambourg. Paradoxes et représentations contradictoires d'une métropole populaire. Bull l'Assoc Géograp Français 93(2), pp.115–136
13. Herzog P (2002) L'Europe après l'Europe. Les voies d'une metamorphose. De Boeck Supérieur, Hors collection, Paris
14. Wolton T (2017) Histoire mondiale du communisme: les complices. Grasset, Paris
15. Schmid JR (1971) [1936], Le maître camarade et la pédagogie libertaire. Maspero, Paris
16. Depraz S (2016a) Économie sociale de marché ou économie verte de marché ? L'équilibre délicat de la durabilité territoriale allemande. Bull l'Assoc Géograp Français 93(1):3–25
17. Lefebvre H (2000) Espace et politique. Economica, Paris
18. Kerste B (2013) Ville entrepreneuriale, ville créative, ville contestée : Hambourg entre 1983 et 2013. Faire Savoir 10:9–18
19. Weinachter M (2007) Hambourg, métropole portuaire international. Regards sur l'économie allemande, 81. En ligne. http://rea.revues.org/654. Accessed on 10 Dec 2017

20. Läpple D (2010) Enjeux des métropoles et IBA Hambourg, Ateliers européens grands projets urbains Paris—Hambourg. Hambourg City University, Support de communication
21. Le Havre Développement (Comité d'expansion économique de la région havraise) (2010) Benchmark des ports nord européens, En ligne. https://www.logistique-seine-normandie.com/benchmark-des-ports-nord-europeens. Accessed on 15 Oct 2017
22. Notteboom TE (2010) Concentration and the formation of multi port gateway regions in the European container port system: an update. J Trans Geograp 18:567–583
23. Lowe K (2015) Inferno. La dévastation de Hambourg, 1943. Perrin, Paris
24. Mazière B (2009) Penser et aménager les agglomérations urbaines : quelques exemples de métropoles européennes. Annales des Mines 4(56), 72–79
25. Dangschat JS, Alisch M (1995) « La nouvelle pauvreté et les stratégies en faveur des quartiers défavorisés à Hambourg. Espace Popul Soc 3:291–303
26. Conseil de l'Europe (1993) Les grandes villes et leur périphérie: Coopération et gestion coordonnée, Communes et régions d'Europe, 51, Édition du Conseil de l'Europe. Strasbourg
27. Escach N (2016) Les stratégies interterritoriales des municipalités de l'Allemagne Baltique: une transition par les réseaux de ville. Bull l'Assoc Géograph Français 93(1):77–97
28. Frérot O (dir.) (2008) L'IBA Emscher Park. Une démarche innovante de réhabilitation industrielle et urbaine. Agence d'urbanisme de l'agglomération lyonnaise Ed. Lyon.
29. HafenCity Hamburg (2013) Themen, Quartiere, Projekte. HafenCity Hamburg, Hamburg
30. Büscher F (2012) Neue Urbanität und Raumwahrnehmung : Das Leuchtturmprojekt «Ikea» im «Sanierungsgebiet Altona-Altstadt S5 Große Bergstraße/Nobistor» in Hamburg-Altona. Hambourg, HafenCity University Hamburg, Hamburg, Mémoire de géosciences
31. Minton A (2012) Ground Control: fear and happiness in the twenty first century city. Penguin UK, Londres
32. Barbier C (2015) Les métamorphoses du traitement spatial de la question sociale: approche croisée de deux grands projets de renouvellement urbain dans les agglomérations de Lille et de Hambourg. Thèse de doctorat de Sciences Politiques, Paris
33. Nebout M (2013) L'Université de Voisinage, une démarche participative. Urbanisme 390:56–57
34. Birke P (2010) Herrscht hier Banko? Die aktuellen Proteste gegen das Unternehmen Hamburg. Sozial Geschichte Online 3:148–191
35. OCDE (2019) Hamburg metropolitan region OECD territorial reviews. OCDE, Paris, Germany
36. Polèse M (2021) Affordable housing: what to do and what not to do. In: Lessons from Montreal and other places. Communication ICCCASU 2021, Montréal
37. Kerste B (2018) Villes transformées, villes contestées: regards croisés sur des luttes gauches-libertaires à Marseille et à Hambourg. Thèse de doctorat, Aix-Marseille
38. Berque A (2014) La mésologie, pourquoi et pour quoi faire? Presses universitaires de Paris Ouest, Paris
39. Ribault T (2019) Resilience in Fukushima: contribution to a political economy of consent. Alternatives 44(2–4):94–118

Chapter 9
Small Towns Ageing—Searching for Linkages Between Population Processes

Jerzy Bański, Wioletta Kamińska, and Mirosław Mularczyk

9.1 Introduction

Small towns have specific social, economic and cultural features that distinguish them from medium-sized and large urban centres on the one hand, and rural areas on the other. They are a significant component in the settlement structures and perform a number of important economic functions [1–3]. Their relationships with their rural surroundings set local development in terms of both direction and level. Small towns are specific kinds of centres of economic development in rural areas [4], hence the focusing of relevant research on relationships with the countryside, as well as the role played in servicing the rural populace [5–8]. One of the factors that determine the development of small towns is their demographic structure including, above all, the age structure of the inhabitants.

The population potential of the towns is shaped primarily by the natural increase and the net migration balance. These elements exert a differentiated influence on the demographic situation of the settlement units, meaning different strength of influence on the dynamics of the demographic structures, including population ageing. The least advantageous population-related situation characterises the towns, in which, side by side with the decrease of the birth rates, there has been an outflow of the young inhabitants. Migrations of the young and active persons are the evidence of the economic stagnation and lack of development perspectives [9, 10]. The consequence

J. Bański (✉)
Institute of Geography and Spatial Organization, Polish Academy of Sciences, Warsaw, Poland
e-mail: jbanski@twarda.pan.pl

W. Kamińska · M. Mularczyk
Jan Kochanowski University in Kielce, Kielce, Poland
e-mail: wioletta.kaminska@ujk.edu.pl

M. Mularczyk
e-mail: miroslaw.mularczyk@ujk.edu.pl

of the deformation in the age structure and of the weakening of the population potential is constituted by the decrease of demand for goods and services. In effect, the economy shrinks, which deepens yet the negative demographic processes. Hoekveld [11] refers to this as to the "vicious circle", leading to the disappearance of the population in a given settlement unit. Similar, though less complex demographic processes concern the collapse of the rural settlement units [12–15].

Population ageing constitutes a global phenomenon, characterising primarily the wealthier countries. It can be generally assumed that more intensive processes of increase in the share of the elderly in the society are observed in towns in comparison with the rural areas and in the periphery rather than in the central areas [16, 17]. The dominating view in the literature is that this phenomenon is the consequence of two unquestioned achievements of humanity: controlling the birth rates and limiting the premature deaths [18]. The pursuit of economic growth and adequate health care, with simultaneous adaptation to the ageing and shrinking labour force constitutes nowadays one of the biggest challenges for the governments of the majority of countries in the world. Currently, the indicator of the number of persons in productive age per one elderly person is low in many regions of the world, while the forecasts indicate that in the coming decades it will be getting even lower.

The subject of ageing occupies an important place in the research literature and is being analysed from many viewpoints. In quantitative studies attention of the respective authors concentrates mainly on the increase in the share of the elderly in the total population numbers (see, e.g., [19–21]) and on the factors, which shape this process, as well as the consequences for the economy [22–24]. In the qualitative studies, the issues are considered of life quality of the elderly, including those related to health care [25, 26] of the loneliness of the elderly (e.g., [27]), of the policies, oriented at the elderly (e.g. [28]), and so on. The studies undertaken concern diverse spatial scales—from local to global [29, 30].

Among the factors shaping the population ageing processes, the most frequently quoted ones are the increase in the standards of living, improvement in sanitary conditions, developments of medicine, birth control, as well as curbing of premature mortality [31, 32]. Paradoxically, the usually negatively assessed phenomenon of population ageing—from the point of view of the socio-economic development of a definite area—is the effect of the highly advantageous and commonly expected circumstances. It was already almost eighty years ago that [33] indicated that the perception of the disadvantageous consequences of the phenomenon in question constitutes a pessimistic way of looking at the great triumph of civilisation. The ageing of the society, in the opinion of [34] is not a problem in itself—the problem is generated by its dynamics and intensity. Namely, the excessive increase in the number, or share, of the elderly leads to the deformation of the age structure of the population, which might exert a negative influence on the state of the economy, the social structure, labour market, services, and so on. Currently, the indicator of the number of persons in productive age per one elderly person is quite low, and the forecasts indicate that in the coming decades it will be yet decreasing. In the years 1990–2017 the share of the elderly (persons aged 60 and more) increased from 9.2 to 14%, and there are forecasts that by 2050 it will reach 21% [35].

Poland is one of the countries, in which the change of the age structure of the population is particularly acute. From one of the youngest societies in the European Union at the beginning of the twenty-first century, Poland is turning into one of the oldest. According to [36], this process is unavoidable and it results, on the one hand, from the change in the childbearing models of the Polish families, and, on the other hand, from the extension of life expectancy. Young families often prefer the 2 + 1 model, putting frequently at the first place the professional career, and delaying the decision of having a child to the later periods. This is accompanied by the changes in cultural conditioning, generating, in particular, the conflict between the traditional and the modern perceptions of the social role of women. These are especially well pronounced on the line countryside-city [37]. In addition, the increased mobility of the Polish society after 1989 caused a significant outflow of young people abroad, which also diminished the childbearing potential.

An interesting research issue is constituted by the demographic situation of the small towns (those with less than 20 000 inhabitants) in Poland. They constitute the biggest group in the structure of the urban settlement network and fulfil numerous essential social and economic functions [38]. These centres are an important link in the system of connections between the countryside and the cities of regional and national significance [5, 39]. The directions of the demographic changes, having taken place over the last several decades, were the resultant of a variety of causes. So, for instance, the small towns, in which large industrial plants were located, or those neighbouring upon the large urban centres, enjoyed altogether a demographic flourishing. On the other hand, those towns that were located in the peripheries or were devoid of the significant economic functions, went through population regression or at most stagnation. It should also be emphasised that in the period of the real socialism the development—including demographic development—of the small towns was heavily influenced by the decisions of the respective authorities. This applies, in a particular manner, to the period of reconstruction of the country after the war damages and the initial period of the socialist economy. Later on, i.e., in the 1970s and in the 1980s, notwithstanding the ambitious Plan of the spatial organisation of Poland till 1990, with the so-called system of moderate polycentric concentration, which assumed, in particular, the development of the industrial functions in the smaller urban centres, the social and economic problems of the country slowed down the investments, which could have propelled the development of these towns (see [40]). In the period of systemic transformation, the local self-governmental authorities undertook the attempts of redefining the role and place of the small towns in the spatial structures of the regions, searching for the new development of impulses [41].

The purpose of the paper is to determine the relations between the natural increase and the net migration balance in small towns and to assess their impact on the contemporary population ageing processes in such settlement units in Poland. The value added of the study, reported here, results from the diagnosis of the population ageing processes in small towns in Poland at the beginning of the twenty-first century and the identification of the main factors, shaping these processes. The results of the investigations here reported may have a bearing for other countries of Central and

Eastern Europe, having a similar conditioning of the socio-economic development [42–44].

9.2 Materials and Methods

According to the data for 2017 there were in Poland 705 small towns. In view of accessibility of the statistical material and the timing of the study (period 2008–2017), the analysis accounted for 670 small towns (i.e., 72.6% of all the towns in Poland). The basic set of statistical data was acquired from the Statistics Poland.

In the first phase of the study the relations were determined between the natural increase and the net migration balance in the towns considered. In order to avoid the influence of incidental annual changes the indicators of the natural increase and the net migration balance for the initial and terminal periods of analysis were averaged over three consecutive years, i.e., 2008–2009–2010 (2008/2010) and 2015–2016–2017 (2015/2017). Eight demographic categories of small towns were distinguished, designated with capital letters from A to H (see Table 9.2), these categories featuring various directions of relations between natural increase and net migration balance (Fig. 9.1). In general terms, the first four categories (A, B, C, D) refer to the units, in which the population number increased, while the subsequent four categories (E, F, G, H)—to towns, in which population numbers decreased (Table 9.1).

The subsequent step consisted in determination of the rate of population ageing in the small towns using the demographic old age indicator (DOA), which is based on the percentage point differences between the shares of the young and old population [46]. The investigations, concerning the age structure of the population, often use the indicator based on the share of the elderly population, which attracts attention uniquely to the persons in post-productive age (65+). It appears that the here applied indicator DOA is more appropriate, as it addresses the relation between the elderly and the young segments of the population in a particular town. The higher the value of this indicator, the more advanced the ageing process of the population. The indicator value was calculated from the formula

$$\text{DOA} = [P(0-14)t - P(0-14)t + n] + [P(\geq 65)t + n - P(\geq 65)t]$$

where:

$P(0-14)t$: average share of population in the age between 0 and 14 years in 2008/2010;

$P(0-14)t + n$: average share of the population in the age between 0 and 14 years in 2015/2017,

$P(\geq 65)t$: average share of the population in the age of 65 and more years in 2008/2010,

$P \geq 65)t + n$: average share of the population in the age of 65 and more years in 2015/2017.

9 Small Towns Ageing—Searching for Linkages Between Population ...

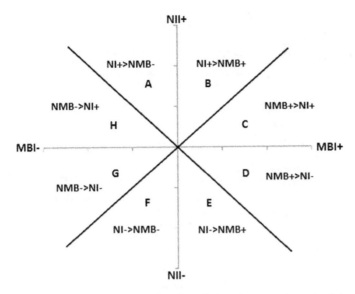

Fig. 9.1 Demographic categories accounting for relations between the natural increase and the net migration balance. A, ..., H—demographic categories of small towns; NI—natural increase; NMB—net migration balance; MBI—migration balance indicator (MBI+ -positive migration balance, MBI—negative migration balance); NII—natural increase indicator (NII+ -natural increase, NII—-atural decrease). *Source* own elaboration on the basis of [45]

Table 9.1 Changes in the demographic types of the small towns in Poland in the period 2008/2010–2015/2017

Demographic types of small towns	2008/2010		2015/2017		Rate and direction of change
	Number	%	Number	%	2008 = 100
A	93	13.9	42	6.3	(−) 54,8
B	51	7.6	24	3.6	(−) 52,9
C	46	6.9	36	5.4	(−) 21,7
D	45	6.7	28	4.2	(−) 37,8
Progressive types total	235	35.1	130	19.4	(−) 46,7
E	35	5.2	39	5.8	(+) 11,4
F	38	5.7	122	18.2	(+) 221,1
G	139	20.7	254	37.9	(+) 82,7
H	223	33.3	125	18.7	(−) 43,9
Regressive types total	435	64.9	540	80.6	(+) 24,1

Source own elaboration on the basis of data from Statistics Poland

Table 9.2 Changes in the number of small towns according to demographic categories in the period 2008/2010–2015/2017

Demographic types 2008/2010	Number of towns	Numbers of towns according to demographic types in 2015/2017							
		A	B	C	D	E	F	G	H
A	93	18	5	2	4	3	5	25	31
B	51	10	10	6	2	0	5	9	9
C	46	6	6	16	4	2	3	5	4
D	45	0	1	4	9	12	10	7	2
E	35	0	0	0	3	5	19	7	1
F	38	0	0	3	0	6	15	12	2
G	139	0	1	0	3	6	52	71	6
H	223	8	1	5	3	5	13	118	70

Source own elaboration on the basis of data from Statistics Poland

On the basis of values of the DOA indicator the towns analysed were classified into four groups[1]:

(1) group Y: DOA ≤ 0, i.e. towns with the population getting younger;
(2) group A1: 0 < DOA < 3.19 (values lower than the mean minus one standard deviation), i.e. towns featuring slow rate of population ageing (below the average);
(3) group A2: 3.20 < DOA < 7.98 (values between the mean minus standard deviation and the mean plus standard deviation), i.e. towns featuring approximately average rate of population ageing;
(4) group A3: DOA ≥ 7.98 (values higher than the mean plus standard deviation), i.e. towns featuring high rate of population ageing (above the average).

Finally, the study of the interrelations between the elements of demographic processes (natural increase, net migration balance) and the rate of ageing of the population made use of Pearson's linear correlation coefficient.

9.3 Analysis and Results

The set of the small towns in Poland, subject to analysis, contained in the initial period of study (2008–2010) the representatives of all the demographic categories, defined before. The regressive urban centres, that is—those with population numbers decreasing, dominated (65% of all towns analysed). The remaining 35% of towns were classified as progressive (see Table 9.1). Then, during the period considered, the share of the regressive towns grew significantly—to 80.6% of all towns analysed.

---

[1] The division into four groups was adopted on the basis of the mean value of the DOA indicator (5.59) and its standard deviation (2.40).

The most numerous among the units of regressive type were the towns, representing the category H, in which the negative net migration balance was not compensated by the natural increase. They accounted for 1/3 of all the units considered. Natural increase ranged in them from 0.01 to 8.1%, while migration balance ranged between −0.6 and −16%. The highest number of such towns was located in the south-eastern and north-western as well as western Poland.

The share of the small centres, representing category H, decreased in the period considered from 33.3 to 18.7%, which indicates the decrease in the significance of the natural processes in the dynamics of population number compared to that of the migration processes (see Table 9.2). The indicators of the natural increase in the towns of this category in the final years of the analysis (2015/2017) ranged from + 0.01 to +6.8%, while net migration balance ranged between −0.8 and −16.6%. The decrease in the share of the centres, belonging to category H, has not changed their spatial distribution. They are still most numerous in south-eastern Poland and in the western part of the country. In the opinion of [47] the inhabitants of the western regions reacted more actively to the social programs, such as 500+ [2] or 300+,[3] and therefrom the childbearing rates above the average. On the other hand, the eastern regions of Poland remain the strongholds of the traditional family and religious values.

The second place in terms of the number of towns in the group considered in the initial period of analysis (some 21% of units accounted for, see Table 9.1) was occupied by the towns representing type G. This type is characterised by the natural decrease and the negative net migration balance, the latter factor being more important for the process of population decrease. The values of the net migration balance in this group of towns ranged from −0.37 to −15.4%, while the natural increase indicator ranged from −0.01 to 8.1%.

The share of the small towns from category G increased over not quite ten years from 20.7 to 37.9% and these centres form now the biggest group of the demographically regressive ones (see Table 9.2). The direct reason for the disadvantageous demographic phenomena is constituted, first of all, by the outflow of young people, to a high extent woman, looking for a better life in the large urban agglomerations. Besides, migrants include a high share of better educated, more socially and economically active persons, which amounts to unfavourable perspectives for the towns, belonging to this category.

The number of towns classified in category F increased during the period considered in a particularly spectacular manner. In these towns natural decrease is greater than the negative migration balance. The share of this category in the total number of towns analysed increased more than threefold, from 5.7% in the years 2008–2010 to 18.2% in the years 2015–2017 (Table 9.2). The common feature of the majority

---

[2] The state-wide program, implemented since April 1st, 2016, of social character, having the purpose of assisting the families in children upbringing through the monthly allotment, disbursed per child in a family of 500 PLN.

[3] The state-wide program, meant to assist the school-going children – constituted by the payment of 300 PLN per pupil at the start of the scholarly year.

of such towns is their peripheral location with respect to the main regional centres. In most cases the towns from category F had belonged, in the initial period of analysis, to category G. This means that the earlier identified large migration loss had a negative impact on the natural processes, which confirms the previously indicated conclusion that it is young people in the childbearing age that most often migrate from the small centres.

It can be generally stated that during the period considered the changes in the groups of towns, classified as regressive (types E, F, G and H) consisted primarily in the transfers between these four categories, the respective towns remaining still in the subset, characterised by population number decrease. Of the 35 units, classified at the beginning of the period analysed in category E, only 5 remained in it at the end of the period. Analogous changes in the remaining categories were as follows: of 38 towns in category F, 15 remained in this category, of 139 towns in category G—71 remained in the same one, and of 223 towns in category H—70 remained in it (Table 9.2).

Among the progressive types of centres the biggest group was constituted in the years 2008–2010 (93 units, 14% of all the towns analysed) by the towns of type A, in which natural increase exceeded the negative net migration. The natural increase indicator ranged in this group of small towns from 0.10 to 7.4%, while the net migration balance ranged from −0.03 to −6.5. The biggest number of small towns of type A was situated in Wielkopolska, then in Carpathian Mts. Migration outflow from the small towns of Wielkopolskie was directed mainly towards the countryside. This outflow, however, was to such an extent small, that it did not exceed the natural increase. On the other hand, in the Carpathian small towns the traditional family and cultural patterns dominate, associated with the high degree of religiosity of the inhabitants.

During the period analysed very significant changes took place in the category A of towns, namely only 18 of them remained in this group. Among those that changed the category, 11 remained in the progressive categories, while the remaining 64 towns had in the final period of analysis the regressive character. A similarly significant reduction took place in the cases of the progressive categories B, C and D. Thus, for instance, in the case of the units, having belonged to category B, their number decreased over the whole period considered from 51 to 10, with 23 towns from this category having moved to the regressive categories (Table 9.2).

The analysis of changes in the demographic categories of small towns in the years 2008/2010–2015/2017 clearly indicates the decrease of population numbers in this group of centres, resulting both from the natural decrease and from migration outflow. The percentage share of the centres considered, featuring negative net migration balance increased in the years analysed from 74 to 81%, while the share of those featuring natural decrease grew from 38 to 66%. There has been more than twofold increase (from 26 to 56%) in the share of small towns, simultaneously featuring negative net migration balance and natural decrease. There has been a distinct drop– from 15% down to 9%–of the share of towns, characterised by the positive values of these two indicators.

In the years 2008/2010–2015/2017 in almost all small towns in Poland the process of population ageing has been observed. This is, definitely, a persistent tendency since the turn of the twenty-first century [48]. The averaged indicator of demographic old age (DOA) for the entire collection of 670 towns considered was equal 5.6 and ranged for this collection from −10.4 to 16.1. A vast majority of small towns, namely 660 out of 670, is characterised by the population ageing processes, taking place at various rates.

In the group A1, in which these processes are slower than on the average, 80 towns were classified. The approximately average rate of population ageing (corresponding to group A2) characterised more than 70% of the towns considered, the value of the old age indicator ranging in these towns from 3.2 to 7.98. Finally, group A3, characterised by the rate of population ageing beyond the average, was composed of 95 small towns (see Table 9.2).

It was assumed in the study here reported that the identified demographic situation of small towns at the beginning of the period considered (i.e., according to the averaged values of indicators for the years 2008–2010) shapes the rate of population ageing of the respective inhabitant populations in the period 2008/2010–2015/2017. The value of Pearson correlation coefficient for the entire set of 670 urban centres considered between the natural increase indicator and the rate of population ageing was at −0.075 (at the significance level of $p = 0.05$). This means that, in fact, the rates of population ageing have not been correlated with the natural increase indicators (Fig. 9.2).

On the other hand, though, the value of the Pearson linear correlation coefficient between the net migration balance indicator and the rate of population ageing was − 0.480 and was statistically significant at $p = 0.05$. Hence, it can be stated that the rate of ageing of the inhabitants, residing in small towns in the years 2008/2010–2015/ 2017 was correlated with the net migration balance. The bigger the net migration outflows, the higher the population ageing rates.

9.4 Discussion and Attempts to Interpret Processes

Taking into account natural increase and the balance of migration, two categories of small towns–H and G–are particularly interesting. The demographic situation of towns category H resulted from a variety of historical, cultural and economic circumstances. The inhabitants of the south-eastern regions are characterised by attachment to religion and to family traditions. In the period of the nineteenth century partitions of Poland this area belonged administratively to two different systems (Austrian Galicia and Russian Congress Kingdom), and this division, as demonstrated by the respective analyses, can be traced in the demographic phenomena even up till now (see [47, 49]). The indicators of fertility and childbearing in the south-eastern regions were higher than the national averages [50]. Yet, the development perspectives, which have been more limited in comparison with other regions of the country (as confirmed by the relatively low socio-economic indicators), motivated the inhabitants of the small

Fig. 9.2 Distribution of small towns according to the rates of ageing of inhabitants in the years 2008/2010–2015/2017 (P-population rejuvenation; R1-below-average aging, R2-average aging, R3-above-average aging). *Source* own elaboration on the basis of data from GUS

towns to undertake migration towards the big urban centres or abroad. The group of migrants was clearly dominated by the young and active persons. On the other hand, the north-western and western regions had been under Prussian influence and after the World War II they were settled by the Polish rural population from the East (mainly from the present-day Ukraine). These population movements, forced by the post-war decisions, resulted in the relatively young demographic structure, which is until today characterised by the above average childbearing indicators. This phenomenon, though, does not compensate for the large migration outflow.

The towns, which belonged to category H, were characterised by the high migration outflow, resulting from the social and demographic conditions. Investigations demonstrate that the actual migration outflows have been much higher than those reported in the official statistics [51, 52]. The factor, stimulating the high level of migration outflow, was constituted by the situation on the labour market after 1989. Only few towns succeeded, after the crisis, caused by the systemic transformation, in developing new economic functions and diversifying the labour market. In the monofunctional towns (mainly factory towns) liquidation or restructuring of the enterprise, being often the sole significant employer, brought crisis on the local labour market and increase of unemployment, which, in turn, constituted an impulse for the inhabitants to leave the place [53]. Decrease of population number (especially of those in productive age) generated other problems, including, in particular, a drop in the revenues of the urban budget, related to the taxes, including local taxes. The economic problems were yet amplified by the disadvantageous social phenomena (joblessness, social exclusion, pathologies, etc.), and lack of interest from the side of the potential investors. The small towns from the category H are also frequently characterised by the unfavourable geographical location. The majority of them is situated within the peripheral areas, at bigger distances from the main regional centres and transport routes.

In the small towns, located in the northern and western regions, the outflow of population resulted also from the collapse of the state farms. In the period of the socialist, centrally managed economy these urban centres were economically activated by the development of the agricultural and food processing industries and fulfilled the service functions for the socialised farming sector as well as for the inhabitants of the surrounding rural areas [54]. Their economic situation changed in the period of economic transformation, with the most disadvantageous changes resulting from the liquidation of the state farms, functioning in the neighbourhoods of these towns. Unemployment surged first of all in the centres, which had been the service nodes for the socialised farming and this became one of the main reasons for the job-related migrations to Germany.

Yet other factors played an important role in the towns, classified in this category, which are located in the Silesia region. In this case, a significant factor in the migration outflow was the fact that an important share of the population was constituted by the persons of German or 'Silesian' extraction. These ethnic groups had bigger possibilities of finding employment in Germany, and hence foreign migrations constituted an important factor in the shrinking of the population resources [51]. Similar circumstances existed also in some of the towns, located in Carpathian mountainous, from where outflow of a part of population took place to the United States.

In turn, the towns representing type G were concentrated in Sudety Mts. and in the north part of the country. In Sudety Mts. the process of depopulation was observed already at the end of the nineteenth century and it was associated with the outflow of population from the mountainous areas to large industrial cities [55]. The second wave of depopulation of the territory of Sudety Mts. was linked with the exchange of population following the World War II, when the resettled German population was replaced by the incoming Polish population, originating from the present-day Ukraine and from south-eastern Poland [56]. Then, in the 1960s and 1970s, the decrease of the population numbers resulted from the fact that the localities, situated at higher altitudes were abandoned by the post-war settlers, discouraged by the difficulty of farming in mountainous areas [57]. In some towns even the well-developed tourism functions and non-market services have not stopped migration outflow.

In the north of the country a significant migration outflow from type G centres was mainly caused by the economic factors and the peripheral location. In the case of Zachodniopomorskie province, which is adjacent to the border with Germany, this location facilitated and motivated to the job-related migration [58]. On the other hand, the economic structure of Warmińsko-Mazurskie province had been dominated by the agricultural function, primarily in the form of the state farms. The inhabitants of the small urban centres were often employed in these state farms. In effect of liquidation of these farms during the 1990s, a large group of employees lost jobs and unemployment soared. Weakly developed network of bigger towns, problems with access to higher-order service, as well as poor resources on the labour market motivated the inhabitants, and mainly the younger ones, to searching for the more attractive places of residence.

The negative net migration balance in the small towns categorised in type G was reflected in the birth numbers. It was namely so that young women, in the childbearing age, as well as, more generally, population in productive age, have been characterised by higher mobility. Outflow of this category of inhabitants brought the decrease in the childbearing indicators. This phenomenon was additionally strengthened by the changing cultural patterns, including those concerning fertility [47, 59].

The results, obtained in this study, confirm the fact of increased intensity of the population ageing processes in Polish small towns. This will have in the future a negative influence on the state of economy and the social structures in small towns. The study by [10] indicates that the new businesses are being established first of all by the young people, who are the main group, leaving the small centres. Ageing of a society means also an intensive erosion of the quality of labour force and social capital in this society. The negative consequence of the migration outflow is constituted by the decrease in demand for goods and services, which entails shrinking of entrepreneurship base and limitation to the use of social and economic infrastructure, leading to the potential liquidation of slack businesses and organisations. The situation arising is translated into the decreased supply of jobs, increase of unemployment and triggering off a new wave of outmigration. Breaking away from this vicious cycle constitutes a true challenge for the self-governmental authorities, because the persistence of a significant migration outflow from small towns entails the succession of the negative consequences of both social and economic nature. These consequences concern not only the small urban centres, but also the rural areas, associated and neighbouring with them.

In the group of towns, characterised by the rejuvenating age structure of the inhabitant population (group Y) only 10 centres were classified. Five of these centres are located in the vicinity of large cities. It can therefore be expected that their development is largely dependent upon the suburbanisation processes. Young people or young families migrate to them both from the regional centres and from the countryside, searching for the lower living costs than in the large cities and, at the same time, better housing conditions [60]. In the majority of small towns, situated close to the regional centres, dynamic growth of family house construction has been observed since the 1990s. The increase in the number of private cars and the improvement in road infrastructure ensure favourable conditions for the daily job commuting. Besides, the flow of young migrants is being attracted to the small towns by the offer of the relatively cheap space for the conduct of own business, lower prices of municipality services, as well as favourable attitude of the local self-governmental authorities to entrepreneurship. The factors listed contribute to the young demographic structure in such centres. The remaining five towns in this group include those having spa or health resort, industrial and transport functions. These centres attract young people with their ample offer on the labour market and convenient road connections with large cities.

Considering the conditioning for the slow rate of population ageing one in the group A1 can distinguish three categories of these towns. The first of them is constituted by the towns located in the shade of the large cities, having supra-regional significance (near Poznan, Warsaw, or Cracow). Location rent and functional connections with the central place, displaying a rich offer on the labour market have most probably resulted in the limitation to the negative changes in the demographic structure. The towns here commented upon were characterised by the positive net migration balance, and the incoming population in productive age slowed down the population ageing processes. The faster rate of ageing than in group P was due primarily to the low childbearing rates. According to the investigations of [59] the inhabitants of the suburban areas took over the new fertility and family models faster than on the areas located farther away from the main economic centres, and this exerted definite influence in terms of the decline in the childbearing indicator. The second category in group A1 is constituted by the units located in the regions, which are "procreation-wise active", with families having many children and with high degree of religious attachment (eastern regions and the Carpathians). Here, the slower rate of population ageing was mainly the effect of natural increase and the fertility levels, exceeding the average. In only few towns natural increase was accompanied by the positive net migration balance, resulting from the labour market offer and convenient location with respect to transport routes. The third category in this group was composed of the settlement centres, which are located in the peripheries, but which are characterised by the dynamic economic development and well developed transport infrastructure. The possibility of daily job and school commuting, along with good living conditions resulted in high childbearing indicators.

In group A2 we can distinguish also in this group several categories of towns in terms of conditions, shaping the population ageing processes. The first such category is constituted by the towns, featuring both low birth rates and excess outflow of population. These towns accounted for 56% (268 units) of the total in group A2. An important proportion of these towns is situated in the so-called problem areas, characterised by the shrinking population potential and delayed economic development [61, 62]. Population ageing was diagnosed there already in the period of socialism, and it intensified yet after the systemic transformation. The second category consisted of the towns (125 units, which featured during the years 2008/2010–2015/2017 only one of the two negative phenomena (natural decrease or negative net migration balance, but with a high intensity. These units are situated in the transitory zones between the large urban centres and the peripheries of the provinces. Population loss was associated in their case primarily with the economic crisis, that is—the liquidation or restructuring of the local industrial enterprises. Liquidation of jobs caused increase of unemployment, and this, in turn, stimulated the inhabitants to leave. In only some of the centres it was possible to maintain the manufacturing functions, but the scale of employment in the local enterprises has been mostly diminished. Similar tendencies were observed also in other post-socialist countries of Central Europe [63, 64]. The third category of these towns featured positive net migration balance and increase of population numbers, but, nevertheless, population ageing has been taking place in these centres, as well. This category contained, first of all, the towns with well-developed tourist functions. Such towns were often chosen as places of residence by the elderly. The factors, attracting such people to these towns were constituted by the attractive natural environment and adequately developed health care services.

The towns representing group A3 are located primarily in north-western and in eastern Poland. In the case of the north-western areas population ageing was due both to the migration outflow and to the low birth rates. These areas have been considered over the post-war decades as demographically young, but at the end of the twentieth century an excess migration outflow from there could be observed. Migrations intensified after the crisis in the agricultural sector, caused by the collapse of the state farm system, and then after the accession of Poland to the European Union. Even the very high tourist attractiveness of some small towns and the seacoast location have not alleviated the demographic problems.

9.5 Conclusions

The first stage of investigations, consisting in the analysis of the indicators of natural increase and net migration balance, allowed for distinguishing eight categories of small towns. The set of towns analysed was dominated by the centres characterised by the significant migration outflow and/or natural decrease. In the period, adopted as the starting point of the study (2008/2010) the dominating category was constituted by the towns, in which the negative net migration balance was not compensated by the positive natural increase (category H). Then, in the final years of the study (2015–2017) the highest proportion of small towns considered belonged to category G, featuring high migration outflow and natural decrease. The share of towns, representing the progressive demographic categories, decreased over the entire period considered.

The study demonstrated that the natural and migration processes in small towns had predominantly negative directions, leading to the weakened population potential, and the direct effect was depopulation. In the initial period of the study 62% of the small towns featured natural increase, while in the final period the towns with natural decrease dominated (67%). Similar changes apply to the changes, concerning migrations. The share of towns, characterised by the net migration outflow increased over the whole period analysed from 74 to 81%.

Further, the share of towns, in which population ageing was observed increased in the years 2008/2010–2015/2017 from 65% to more than 80%. The population ageing process was taking place in almost all of the small towns in Poland, and there were only 10 of them, in which the age structure of the resident population was getting younger. The biggest changes were occurring in the small towns, situated in north-western and central-eastern Poland. These were, first of all, the centres featuring high migration outflows. The slowest rates of ageing were observed in the small centres situated in the southern belt of provinces.

The analysis of interdependences between the natural increase indicator and the net migration balance on the one hand and the old age indicator on the other demonstrated that the course of the population ageing processes in small towns depends primarily upon the contemporary directions of migrations. No significant influence was identified, though, of the natural increase (or decrease) on the rate of population ageing. It appears to be necessary to turn the attention of the self-governmental authorities to the need of carrying out an adequate policy, aiming at the limitation of the consequences of the population shrinking and ageing in small towns in Poland. Given the number of these centres, their significance in the settlement network and spatial organisation of the country is high. The disadvantageous changes in the population potential, which take place in these urban centres may have negative consequences for the entire regions.

Acknowledgements Publication prepared under the research projects of the National Science Centre, nb. UMO-2019/35/B/HS4/00114, *Diagnosis of the contemporary socio-economic structure and functional classification of small towns in Poland - in search of model solutions.*

References

1. Cox E, Longlands S (2016) City systems. The role of small and medium-sized towns and cities in growing the northern powerhouse. Institute for Public Policy Research North, Sprinningfields
2. Hopkins J, Copus A (2018) Definitions, measurement approaches and typologies of rural areas and small towns: a review, Report. The James Hutton Institute, Aberdeen
3. Meijers, E., Cardaso, R. (2021) Shedding light or casting shadows? Relations between primary and secondary cities. In: Pendras M, Williams C (eds.) Secondary cities: exploring uneven development in dynamic urban regions of the global North. Bristol University Press, pp 25–54
4. Shucksmith M, Thompson KJ, Roberts D (2005) The CAP and the regions. In: The territorial impact of the common agricultural policy. CABI Publishing, Wallingford
5. Dej M, Janas K, Wolski O (Eds) (2014) Towards urban-rural partnership in Poland. In: Preconditions and potential. Institute of Urban Development, Kraków
6. Edwards B, Goodwin M, Woods M (2003) Citizenship, community and participation in small towns: a case study of regeneration partnerships. In: Imrie R, Raco M (eds) Urban renaissance? New labour, community, and urban policy. Policy Press, Bristol, pp 181–204
7. Powe NA (2013) Market towns: roles, challenges and prospects. Routledge, Abingdon
8. Van Leeuwen E (2010) Urban-rural interactions: towns and focus points in rural development. Springer Science and Business Media, Berlin
9. Bański J, Flaga M (2013) The areas of unfavourable demographic processes in Eastern Poland—selected aspects. Barometr Regionalny 11(2):17–24
10. Kamińska W (2006) Pozarolnicza indywidualna działalność gospodarcza w Polsce w latach 1988–2003. Prace Geograficzne, IGiPZ PAN, Warszawa, p 203
11. Hoekveld JJ (2012) Time-space relations and the differences between shrinking cities. Built Environ 38(2):179–195
12. Bański J, Wesołowska M (2020) Disappearing villages in Poland—selected socioeconomic processes and spatial phenomena. Europ Countryside 12(2):222–241
13. Collantes F, Pinilla V (2011) Peaceful surrender: The depopulation of rural Spain in the twentieth century. Cambridge Scholars Publishing, Newcastle-upon-Tyne
14. Di Figlia L (2016) Turnaround: abandoned villages, from discarded elements of modern Italian society to possible resources. Int Plan Stud 21(3):278–297

15. Dyer C, Jones R (eds) (2010) Deserted villages revisited, hatfield
16. Hoff A (ed) (2011) Population ageing in central and Eastern Europe: societal and policy implications. Ashgate Publishing
17. Rosenbers MW, Wilson K (2018) Population geographies of older people. In: Skinner WMW, Andrews GJ, Cutchin MP (red.)(eds.), Geographical gerontology. perspectives, concepts, approaches. Routledge, London, pp 56–67
18. Lorenti A (2015) Investing on ageing. Extending working lives across Europe. Sapienza Universita di Roma, Roma. http://padis.uniroma1.it/bitstream/10805/2719/1/thesis.alorenti.pdf. Accessed on 3 April 2021
19. Anderson GF, Hussey PS (2000) Population aging: a comparison among industrialized countries. Health Aff 19(3):191–203
20. Mustafina M (2020) Quantum and tempo effects of changes during the demographic transition: classification of world sub-regions and selected countries. AUC Geograp 55(1):15–26
21. Sanderson WC, Scherbov S (2007) A new perspective on population ageing. Demogr Res 16(2):27–58
22. Batljan I, Lagergren M, Thorslund M (2009) Population ageing in Sweden: the effect of change in educational composition on the future number of older people suffering severe ill health. Eur J Ageing 6:201–211
23. Bloom DE, Canning D, Fink G (2011) Implications of population aging for economic growth. In: NBER Working Paper Series, 16705. National Bureau of Economic Research, Inc., Cambridge
24. Dixon S (2003) (2003) Implications of population ageing for the labour market. Lab Market Trends 111(2):67–76
25. Fried LP, Paccaud F (2010) Editorial: the public health needs for an ageing society. Public Health Rev 32(2):351–355
26. Rosenthal L (2009) The role of local government: land-use controls and aging-friendliness. Generations 2:18–25
27. Atchley RC (1976) The sociology of retirement. OECD, New York
28. Lena A, Ashok K, Padma M, Kamath V, Kamath A (2009) Health and social problems of the elderly: a cross-sectional study in Udupi Taluk, Karnataka, Indian. J Comm Med 34:131–134
29. Bengtsson T, Scott K (2010) The ageing population. In: Bengtsson T (ed.) Population ageing– a threat to the welfare state? Demographic research monographs. Springer-Verlag, Berlin, Heidelberg. https://doi.org/10.1007/978-3-642-12612-3_2
30. Sinigoj G, Jones G, Hirokawa K, Linhart S (eds.) (2007) The impact of ageing. A common challenge for Europe and Asia. Forschung und Wissenschaft, Kultur
31. Crampton A (2009) Global aging: emerging challenges, The Pardee Papers, 6. Boston University, Boston
32. Palacios R (2002) The future of global ageing. Int J Epidemiol 31:786–791
33. Notestein FW (1945) Population the long view. In: Schultz TP (ed) Food for the world. University of Chicago Press, Chicago, pp 36–57
34. Golini A (1997) Demographic trends and ageing in Europe. Prosp Probl Polic Genus 53(3–4):33–74
35. World Population Ageing (2017) Report, department of economic and social affairs population division. United Nations, New York
36. Okólski M (2010) Wyzwania demograficzne Europy i Polski (Demographic challenges for Europe and for Poland; in Polish). Stud Socjol 4(199):37–78
37. Bański J (2017) Rozwój obszarów wiejskich (Development of rural areas; in Polish). PWE, Warszawa
38. Bański J (2021) The functions and local linkages of small towns—a review of selected classifications and approaches to research. In: Bański J (ed) The Routledge handbook of small towns. Routledge, New York- London, pp 7–19
39. Kamińska W, Mularczyk M (2014) Demographic types of small cities in Poland. Miscellanea Geogr 18(4):24–33

40. Malisz B (1984) Trzy wizje polskiej przestrzeni (Three visions of Polish space; in Polish). In: Kukliński A (ed.) Gospodarka przestrzenna Polski. Diagnoza—rekonstrukcja—prognoza, vol 125. Biuletyn KPZK PAN, Warszawa, pp 6–45
41. Heffner K (2008) Funkcjonowanie miast małych w systemie osadniczym Polski w perspektywie 2033 r. (Functioning of the small towns in the settlement system of Poland in the perspective of the year 2033; in Polish). In: Saganowski K, Zagrzejewska-Fiedorowicz M, Żuber P (eds.) Ekspertyzy do Koncepcji Przestrzennego Zagospodarowania Kraju 2008–2033, vol 1. Ministerstwo Rozwoju Regionalnego, Warszawa, pp 281–333
42. Novotný L, Csachová S, Kulla M, Nestorová-Dická J, Pregi L (2016) Development trajectories of small towns in East Slovakia. Europ Countryside 8(4):373–394
43. Sykora L, Bouzarovski S (2012) Multipe transformations: conceptualising the post-communist urban transition. Urban Stud 49(1):43–60
44. Zamfir D (2007) Geodemography of the small towns in Romania. Universitary Publishing House, Bucharest
45. Jagielski A (1978) Geografia ludności (Geography of population; in Polish). PWN, Warszawa
46. Długosz Z (1997) Stan i dynamika starzenia się ludności Polski. Czasopismo Geograficzne 68:227–232
47. Szukalski P., (2019) Demograficzne ślady zaborów (Demographic traces of the partitions of Poland; in Polish). http://www.sprawynauki.edu.pl/archiwum/dzialy-wyd-elektron/323-demografia-el/4059-demograficzne-slady-zaborow. Accessed on 10 April 2022
48. Kwiatek-Sołtys A (2006) Migration attractivenes of small towns in the Małopolska Province. Bull Geogr Soc Econ Ser 5:155–160
49. Rosner A (ed.) (2007) Zróżnicowanie poziomu rozwoju społeczno-gospodarczego obszarów wiejskich a zróżnicowanie dynamiki przemian (The differentiation of the levels of socio-economic development in rural areas vs. the differentiation of the transformation dynamics; in Polish). Instytut Rozwoju Wsi i Rolnictwa PAN, Warszawa
50. Szukalski P (2015) Przestrzenne zróżnicowanie dzietności w Polsce (Spatial differentiation of child bearing in Poland; in Polish). Wiadomości Statystyczne 4:13–27
51. Solga B (2013) Miejsce i znaczenie migracji zagranicznych w rozwoju regionalnym (The place and the role of foreign migrations in the regional development; in Polish). Instytut Śląski, Opole
52. Śleszyński P (2018) Mapa przestrzennego zróżnicowania współczesnych procesów demograficznych w Polsce (Map of spatial differentiation of the contemporary demographic processes in Poland; in Polish). In: Hrynkiewicz J, Witkowski J, Potrykowska A (eds.) Sytuacja demograficzna Polski jako wyzwanie dla polityki społecznej i gospodarczej. Rządowa Rada Ludnościowa, Warszawa, pp 84–108
53. Kantor-Pietraga I, Krzysztofik R, Runge J (2012) Kontekst geograficzny i funkcjonalny kurczenia się małych miast w Polsce południowej (The geographic and functional context of shrinking of the small towns in southern Poland; in Polish). In: Heffner K, Halama A (eds.) Ewolucja funkcji małych miast w Polsce. Studia Ekonomiczne, Zeszyty Naukowe Wydziałowe, pp 9–24
54. Jażewicz I (2005) Przemiany społeczno-demograficzne i gospodarcze w małych miastach Pomorza Środkowego w okresie transformacji gospodarczej (Socio-demographic and economic transformations in the small towns of Middle Pomerania in the period of economic transformation; in Polish). Słupskie Prace Geograficzne 2:71–79
55. Kościk J (1990) Przemiany demograficzno-osadnicze na Ziemi Kłodzkiej w XIX w. (Demographic and settlement transformations in the Land of Kłodzko during the 19th century; in Polish), Acta Universitatis Wratislaviensis, 832. Historia 53:85–98
56. Jerkiewicz A (1983) Wybrane problemy ludnościowe i osadnicze w Sudetach (Selected demographic and settlement problems in Sudety Mts.; in Polish), Acta Universitatis Wratislaviensis, 506. Studia Geograficzne 32:11–21
57. Latocha A (2013) Wyludnione wsie w Sudetach. I co dalej? (Depopulated villages in Sudety Mts. And what next? In Polish). Przegląd Geograficzny 85:373–396
58. Horolets A, Lesińska M, Okólski M (eds.) (2018) Raport o stanie badań nad migracjami w Polsce po 1989 roku (Report on the state of research on migrations in Poland after 1989; in Polish). Komitet Badań nad Migracjami PAN, Warszawa

59. Kurek S, Lange M (2013) Zmiany zachowań prokreacyjnych w Polsce w ujęciu przestrzennym (Changes in the childbearing behaviour patterns in Poland in the spatial perspective; in Polish), Prace Monograficzne, 658. Uniwersytet Pedagogiczny, Kraków
60. Zborowski A, Raźniak P (2013) Suburbanizacja rezydencjonalna w Polsce–ocena procesu (Residential suburbanisation in Poland–an assessment of the process; in Polish). Studia Miejskie 9:37–50
61. Bański J (2002) Typy ludnościowych obszarów problemowych (Types of the demographic problem areas; in Polish). In: Bański J, Rydz E (eds) Społeczne problemy wsi, Studia Obszarów Wiejskich, 2, PTG. IGiPZ PAN, Warszawa, pp 41–52
62. Eberhardt P (1989) Regiony wyludniające się w Polsce (Depopulating regions in Poland; in Polish). Prace Geograficzne, IGiPZ PAN, Warszawa, p 148
63. Ianoş I (2016) Causal relationships between economic dynamics and migration. Romania as case study'. In: Dominguez-Mujica J (ed) Global change and human mobility, series title: advances geographical and environmental sciences. Springer, pp 249–264
64. Pirisi G, Trócsányi A (2014) Shrinking small towns in Hungary: the factors behind the urban decline in "small scale." Acta Geograph Univers Comenianae 58(2):131–147

Part III
Urban Development from Diverse Perspectives

Chapter 10
China's Integrated Urban–Rural Development: A Development Mode Outside the Planetary Urbanization Paradigm?

Chen Yang and Zhu Qian

10.1 Introduction

In 2007, the United Nations reported for the first time that more than half of the world's population lives in cities, which signifies the advent of the "Urban Age" [1]. The notion of planetary urbanization is first proposed in Lefebrve's writing, indicating an ultimate form of capitalist production, the "complete urbanization of society" [2]. However, the development trajectory of cities worldwide has suggested that complete urbanization is experienced unevenly in cities, especially when it comes to the dichotomy of the Global North and Global South. In many senses, urban practices in the Global South replicate those of the Global North since the latter represents the most advanced approaches to industrialization and urbanization. The western development modes and urban theories have been translated and imposed on countries in the Global South, with scarce attention paid to the vernacular and local knowledge. The consistent effort to decolonize knowledge production has gained momentum in the context of the re-emergence of critical theories such as post-colonialism and post-modernism [3–5]. A similar situation occurs in the youthful field of urban geography. Onset from the Chicago School's urban models in the 1930s and 1940s, the focal point of urban geography has been exclusively cities in the Global North. Only in the late 1990s, with the emerging globalization [6], urban geographers have engaged in innovative theoretical and empirical contributions based in the Global South, albeit limited [7]. The planetary urbanization [8, 9] as a new paradigm shift in urban geography furnishes geographers with an inclusive

---

C. Yang (✉) · Z. Qian
School of Planning, University of Waterloo, Waterloo, ON, Canada
e-mail: c273yang@uwaterloo.ca

Z. Qian
e-mail: z3qian@uwaterloo.ca

theoretical framework to conceptualize the ongoing urbanization processes in Africa, Asia, and Southern America. However, while the notion is inherently Western, it has obscured and ignored some key aspects of urbanization when viewed from the outside [10]. As such, it is essential to add knowledge from the outside to complement the theoretical and epistemological framework.

China's sheer magnitude of population, low per capita GDP, and uneven wealth distribution characterize China as a developing economy. Although some scholars point to China's exceptionalism regarding political clout [11], economic development path [12], and urbanization [13], China is still listed as one of the 78 countries in the Global South according to the United Nations' Finance Center for South-South Cooperation. More importantly, when it comes to China's underdeveloped and lagging rural regions, there is a huge gap between China and the Global North countries. China's development initiatives that aim to restructure urban–rural relations could offer alternative lenses on urbanization and rural development. Although [12] cautions that Chinese models may not possess universal application since it is guided by pragmatism and political distinctiveness, we believe China's lessons are beneficial to other countries in the Global South to reflect on such models to shed light on their own developmental path. In this paper, we regard China as an important member of the Global South but are also aware of the Chinese exceptionalism in examining China's urban–rural integrated development.

The Nobel Laureate Joseph Stiglitz once commented that technological advancement in the U.S. and the urbanization in China are two key successes of humankind in the twenty-first century. Since the foundation of the People's Republic of China in 1949, China has embarked on its "catch-up" road to its western counterpart countries such as the U.S. and U.K. The pre-reform China is dominated by a planned economy borrowed from the Soviet Union where industrialization was prioritized over urbanization. In 1978, Deng Xiaoping reoriented China's development towards a pro-market mode, which in Harvey's view as an embracement of neoliberalism [14]. Since then, China's urbanization rate has soared. In 2011, four years after the semi-urbanization of the world, China also witnessed its semi-urbanization. China's lessons of urbanization should by no means be set aside from the mainstream scholarship of urban studies when it comes to its distinctive political-economic context and sheer population magnitude. As [15] reminded us a long time ago, mainstream scholars should be open to "alternative perspectives that may not fit well with paradigms now dominating Western social sciences" (p. 1548). Therefore, this paper aims to place China's urbanization process in the broad discourse of planetary urbanization to: (1) critically reflect on the theory of planetary urbanization from the outside, particularly China, and (2) introduce China's urban–rural integrated development mode to the conceptual framework of planetary urbanization as a supplement to its theoretical omissions based on the case of Hangzhou. Admittedly, knowledge is inherently biased [16]. This paper is not intended to challenge the planetary urbanization theory harshly but to remind urban scholars to "abandon historically privileged spots for observing urbanization" ([17], p. 21) to realize the intention of holistically examining the urban.

The remainder of this paper is organized as follows. We first engage in the discussion revolving around planetary urbanization, introducing its fundamental arguments and emerging critiques to the thesis. The subsequent section delves into urban–rural relations and the conceptualization of China's integrated development strategy in relation to planetary urbanization. Following such theoretical debates, the case of Hangzhou is presented, representing typical integrated development practices in China. Secondary data sources such as official statistics, news reports, published papers, and research reports from local planning institutes are used to introduce the case. The paper concludes with a conceptual framework of "planetary urbanization in China" to advance further studies that intend to link China's urbanization to the broad discourse of capitalist production.

10.2 The Thesis of Planetary Urbanization

Urban geography is a relatively youthful sub-discipline of human geography. The intellectual battlefield of the urban has stimulated numerous theoretical contributions to urban geography, including but not limited to urban morphology and urban form, gentrification, globalization and network society, social justice and critical urban studies, and new city science. The most up-to-date contribution made to urban geography is the concept of "planetary urbanization" or "Urbanization I" termed by Derickson [9] and its derivative research [18–20]. This new epistemology has pushed forward the field in three recognized ways: (1) to examine the notion of urban rather than the conception of the city, (2) to interpret the notion of urbanization as a liminal process, and (3) to see urbanization as a variegated, center-less process [21]. The term "planetary urbanization" is first proposed by [8] that draws on Henri Lefebvre's notion of "planetary scale" and "complete urbanization of society" [2], and later refined in their critique of the urban age thesis [1]. However, the most comprehensive exposition of the notion came with their widely cited work–*Towards a new epistemology of the urban* [22]. In their formulation, the urban/urbanization should be understood as theoretical categories rather than empirical objects, a process not a universal form, an assemblage of three mutually constitutive moments, a multidimensional fabric, and a planetary process, a variegated and uneven unfolding, and a collective project. The planetary urbanization therefore is not exclusively about the assertion of the scale of "planetary," but more of a timely epistemological framework for urban geographers to accept this epistemological reflexivity in carrying out further research.

However, what makes the theory problematic in its universal application in the Global South given its overemphasis on the urban realm. Brenner and Schmid [22] contended that "the urban has indeed become a worldwide condition in which [...] long-entrenched geographical divisions (urban/rural, city/countryside, North/South) [are enmeshed]" (p. 173). This claim is rooted and nurtured in the soil of western urbanization. Schmid [23] admitted that his departure point is based on Zurich and Lefebvre's observation of Paris. The already developed and highly urbanized context is much appealing to urban geographers for making consensus on a complete urbanized future that bound to happen. Nonetheless, this future may not speak to countries in the Global South, at least in the foreseeable future. Indeed, it is still vital to admit the persistence of the urban–rural duality in the Global South. In other words, urban is meaningless without the rural serving as the foil, or at least meticulous attention should be paid to the grey fields that are malleable and porous under suburbanization and differential urbanization [17, 24]. The planetary urbanization holds that the non-urban realm "has been internalized into the very core of the urbanization process" ([22], p. 174) because the urban/rural distinction has come to obscure. This deliberate downplaying of the non-urban is a conceptual necessity for the new epistemology, but it does not lead to a comprehensive account of ongoing urbanization, especially in the global south. The pervasive resettlement [25], informality [26], rural land issues [27, 28], and the most recent Indian farmers' protest have reminded us of the indelible role of the non-urban realm in unpacking urbanization in developing countries. This urban-centered mindset reminds us of Lipton's notion of urban bias [29] that regards rural areas as inferior to urban areas. Although planetary urbanization does not engage in the discourse of urban–rural relation bluntly, it implies that the rural, not as a spatial unit, is subordinate to urban and is bound to undergo urbanization due to "the diffusion of urbanism" ([22], p. 169). Although we concur with Brenner and Schmid that under the capitalist logic of production, rural regions serve urban areas through urbanization, we are also mindful of China's urban–rural relation shift influenced by the socialist ideology, which warrants further attention. Before delving into China's distinctive urban–rural relations, we will offer a peek into some typical critiques and intensive debates of planetary urbanization.

There are epistemological alternatives to planetary urbanization: planetary suburbanization [17, 20], the paradigm of Southern urbanism [21], the "Urbanization 2" (Derickson [9], and the view from the outside [10, 30]. The frictions are largely due to the concept of planetary urbanization is deeply rooted in the Marxist view of society that has been criticized for being biased and neglecting other postmodernist perspectives [31]. As [32] himself summarized and specified, the primary criticism centers on the following: (1) planetary urbanization neglects the reflexive questions of positionality in social theory, (2) planetary urbanization pays limited or no attention to questions around gender, sexuality, race and difference given its white, masculinist, and Eurocentric tradition; (3) the framework impedes potential productive engagement; and (4) the idea is homogenous, universal, and imperial. In defending their original intention of proposing the notion of planetary urbanization, [23, 32] call on critics to read into the inner logic of the concept of planetary urbanization and to contextualize the concept in both author's intellectual development process. It should be acknowledged that while planetary urbanization has provoked numerous discussions, debates, and controversies, it is still an infant theory that needs consistent innovations from both the inside domain and the confrontations with the outside.

To sum, the concept of planetary urbanization as a buzzword in the field of urban geography has sparked massive debates, which is precisely what Brenner and Schmid had in mind. Indeed, as noted by [32], p. 570), the intention is to form an "engaged pluralism [where] more productive possibility for dialogue among critical urban researchers [are offered]." From this point onwards, this paper is not a critique of planetary urbanization, rather, a dialogue between the notion with China's urbanization praxis. It intends to identify the similarities and dissimilarities between abstract urbanization with the on-ground evidence from China's unique historical and political-economic settings to advance the comprehensiveness of planetary urbanization in the applicability of the theory.

10.3 China's Experience of Urbanization and Urban–Rural Integration

10.3.1 Conceptualizing China's Urbanization

China's urbanization has undergone a series of paradigm shifts [33] where urban–rural relations have been under intensive transformations. Since the Communist Party of China took power in 1949, China's central government had employed the planned economy under socialist ideology. In Maoist China, urban areas were designated as the locomotives for industrialization and modernization which aims to "catch up" developed Western countries. At that time, a nationwide agrarian campaign swept across China, which redistributes farmlands to the poor (impoverished farmers) by confiscating farmlands from the rich (landlords) (Ding, 2003). It was not until 1958 when the People's Commune (PC) was introduced and the national rural production cooperative movement that marks the establishment of collective ownership of China's rural land [34]. Although under the socialist banner of equality, the heightened urban–rural segregation existed due to the socio-economic legacy of the war times of China and the institutional regulations that deliberately restrict rural people's mobility. In this context, the spatial agglomeration and scale economy in Chinese cities rest on the state's intervention rather than economic laws. China remained a poor agricultural country and the urbanization rate increased by just 8% from 1949 (10%) to 1980 (18%). The immature urbanization under the planned economy is far from the shared prosperity of urban and rural areas depicted by socialism. While the central government attempted to mitigate the urban–rural divide through population control measures [35], the urban–rural relationship was made antagonistic to deploy the urban-biased development mode [36].

In post-reform China, the entry of the market economy has unleashed productivity in both urban and rural areas, which substantially promotes economic development and the urbanization rate. China's urbanization rate has soared since the 1980s and reached 50% in 2011 and 60% in 2019. However, the economic boom has not brought about an easing of the urban–rural dichotomy, and the capitalist logic of production has instead exacerbated the urban–rural gap. In addition to the widening urban–rural economic gulf, the concomitant social costs come in various forms, including the migration, resettlement, and displacement of peasants [33]. The creative destruction of social relations resembles more that of a corollary of capitalist urbanization that proliferates in Asia and other regions of the global South [37–39]. The externalities of such social costs sound the alarm of imbalance development in urban and rural areas, which is clearly at odds with the ultimate goal of urban–rural integration under socialism. Since 2002, the "coordinated urban–rural development" was brought to the fore, followed by a series of policy reforms that aim to address the drawback and lacking rural development [40, 41]. Recent policy reorientation towards the revitalization of the countryside further affirms China's commitment to the rural, a notion some labelled "ruralization" [42]. Indeed, we will show in subsequent sections that this development mode does not fall under either urbanization or ruralization, but a function-oriented resources allocation under Chin's authoritarian planning apparatus [43].

Given China's distinct historical and socio-political context, urbanization in China has been reversed several times to exercise the government's intervention to natural market mechanisms. Capitalist urbanization in developed countries is marked by the capital switch and circulation [44], urban agglomeration [45], and the most recent planetary urbanization condition [22]. While in developing countries, the process is characterized by informality [26] and environmental degradation [38] due to the exploitation of land and human resources. The urban–rural divide is internalized through long-term primitive accumulation during the early urbanization process in developed countries but has become problematic for developing countries when facing the surge in capital accumulation in the form of international capital investment. The spatial fix to developed countries' crisis, therefore becomes a spatial disruption for local developing countries. The development-induced resettlement and displacement (DIRD) in Asian countries has been well documented in the existing literature [46, 47]. Due to the socialist ideology, the Chinese state has deliberately balanced developments in rural and urban areas. In history, China implemented various counter-urbanization measures in history, including the People's Commune movement in China [48], the down to the countryside movement [35], the hukou regime [49], the small town enterprises [50], and the rural revitalization [51]. Such measures, especially those enacted in pre-reform China, were state apparatus that bear political imperatives to forcefully create an egalitarian society, which caused indelible social and political turmoil. After the 1978 reform, while deliberately embracing the efficiency of economic development brought by the market economy, China has been watching closely for the trajectory of development, especially that of the rural. To this end, China's urbanization unfolds in a fashion that involves constant recalibrations by the government.

In addition to understanding China's urbanization from the perspective of urban–rural relations, prior literature has documented much about China's urbanization pattern. Socialist China witnessed a society that is sustained by a low urbanization rate, city-based industrialization, and a phase of anti-urbanism under the political turmoil [52]. Although the then China resembles urbanization under the planned economy and the ultra-leftist political ideology in the former Soviet Union (FSU) and East and Central European (ECE), [15] noted that "no general theory of socialist urbanization […] can adequately account for the diverse facets of China's change during the pre-reform era" (p. 1551). The distinctive political-economic and historical context has imprinted in China's path of urbanization. As entering the post-reform era, China's urbanization is characterized by: (1) the state's dominance in economic development and urbanization, and (2) the 'incomplete urbanization' approach that fuels urban development through massive rural migrant workers [53]. The central government even used administrative boundary adjustment to create cities, which is termed "administrative urbanization" [54–56]. The city-leading-county system established in the 1980s and the massive-scale promotion of the conversion from county to city from 1986 contributed to the surging number of county-level cities in the 1990s. However, the 'pseudo-urbanization' [57, 58] where urban development comes in the form of urbanization rather than industrialization has received wide criticism. Since the late 1990s when urban land became tradable, China has embarked on the landed urbanization course [59], which stimulated a large body of literature in this vein [60–63]. In recent years, the discourse on China's urbanization has diversified, and emerging topics such as new urbanization policy [64], small town urbanization [65] or urbanization for rural development [33], and urbanization through resettlement [66] have also added to the holistic understanding of China's urbanization pattern. Besides such contributions, marrying China's experience with western theory can further advance this knowledge.

As we argued above, planetary urbanization is an epistemological framework based in a Western context and capitalist urbanization, it thus may be ineffective in unravelling China's urbanization, at least in predicting the future condition of urbanization. In this sense, we present some of our reflections on planetary urbanization in reference to China's integrated urban–rural development.

10.3.2 Viewing China's Urbanization Through the Perspective of Planetary Urbanization

The seven theses proposed in Brenner and Schmid's seminal work [22] are extremely insightful and constructive to our understanding of China's urbanization. Therefore, we made the following arguments as the departing point to engaging in the discourse. While we agree with most theses and acknowledge their explanatory potential in China's context, we hope to add China's knowledge to expand on Thesis 3 and Thesis 5 since they are well established for capitalist urbanization but inadequately address some current practices and potentials in China to be comprehensive.

"Thesis 3: urbanization involves three mutually constitutive moments—concentrated urbanization, extended urbanization and differential urbanization" ([22], p. 166). Thesis 3 calls for a dynamic interpretation of urbanization as a process consisting of three "dialectically interconnected and mutually constitutive" moments. This assertion reminds us of Lefebvre's three moments of space production, which offer a dialectical exposition of the capitalist production of urban space [67]. Central to the three moments of urbanization is the extended urbanization, which involves the

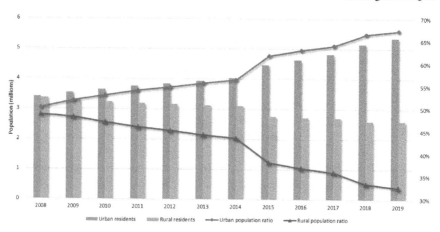

Fig. 10.1 Urban and rural population in Hangzhou, 2008–2019 (*source* [84])

operationalization, reorganization, and enclosure of land far beyond urban centers. Such efforts are devised to support social and economic activities in urban areas but are bound to inflict the creative destruction of socio-spatial relations in non-urban areas. The latter is conceived of as the differential urbanization. It is worth noting that the differential urbanization can also result from urban contradictions immanent to capitalism, such as "class struggle, property relations, over accumulation, and the political control of surplus value" (p. 168). Ultimately, the three moments together shape a planetary urbanization condition where the spatial units, regardless of urban or rural, fall under these three moments and fit their function in the framework (see Fig. 10.1 in [22]).

This framework is relatively comprehensive in dissecting capitalist urbanization, but it may be ineffective when applied to China's urbanization which is shaped by the combining forces of capitalism and socialism. First, we argue that extended urbanization may not serve the agglomeration process in cities. This is evident in China's recent introduction of new countryside in the rural areas under the macro policy rhetoric of poverty alleviation [68–70]. The concomitant massive land consolidation in rural areas bears two primary objectives: (1) providing land for urban development and (2) improving the livelihood of the peasantry. While the former falls under the mechanism of extended urbanization that favors development in urban centers, the latter heed to rural resources, population, and development. This is achieved not through the capital switch in a market economy but by planning intervention in an authoritarian regime. Moreover, the "uneven thickening and stretching of an 'urban fabric'" did not occur, instead, the rural landscape has become more like the countryside rural [71]. Because of this, our second argument is that differential urbanization could lead to new forms of rural space, not only urban space. This is a paradox embedded in planetary urbanization itself. Since the theory presupposes a planetary condition of urbanization, the non-urban outside has been internalized into a dynamic and perpetual process of turning non-urban into urban. However, if the preconditions cannot stand on their own due to the dismantling of its underpinnings of capitalist logic, then the non-urban outside can reverse the process, creating an "urban" outside. The crux of this paradox can be attributed to the subjectivity and situatedness of the new epistemology where it neglects the forces other than capitalism—socialist planning. It is therefore in China, the two forces at play complicate the three moments framework of planetary urbanization. Similarly, it further challenges thesis 5 of the notion.

"Thesis 5: urbanization has become planetary" ([22], p. 172). This thesis can be translated to a more intelligible expression, that is, "capitalist mode of production has become planetary." Brenner and Schmid claimed that the planetary formation of urbanization emerged in the 1980s in the context of a global restructuring that "began with the deconstruction of Fordist-Keynesian and national-developmentalist regimes […] and continued until the withering away of state socialism" (p. 172). It is evident the trending deregulation, globalization, and neoliberalization has contributed much to the proliferation of a homogenous capitalist production worldwide. It is thus the non-urban realm has been made subordinate to the hegemonic discourse of urbanization, and hence the "urban/rural distinction has come to obscure" (p. 175). Brenner and Schmid called for "contextually specific and theoretically reflexive investigations" of the rural areas. To this end, we argue that in China, it is both theoretically and practically untenable to accept that the rural realm has been marginalized and internalized in urbanization. Indeed, rural development has consistently been the priority of China's central government in policy devising and decision making,[1] especially in the recent years after President Xi Jinping proposed Rural revitalization in 2014. Besides, the longstanding urban–rural divide is less likely to reconcile in the short term, given that the state strictly controls the mobility of the rural population and the production elements of rural land [40, 72]. The dichotomy of urban–rural property rights [73–75] and the household registration system (*hukou zhidu*), as well as other institutional arrangements, reaffirm that the urban/rural opposition remains the epistemological anchor for mainstream urban studies in China, at least in the foreseeable future. More importantly, we believe that planetary urbanization has to collaborate with the concept of urban–rural integration in China to achieve a comprehensive and inclusive theorization of development in both urban and rural China.

10.4 Urban–Rural Integration: An Outlier of Planetary Urbanization?

China's urban–rural integration has been examined extensively [33, 76–78], but to the best of our knowledge, none have attempted to facilitate dialogues between this development mode with the hegemonic capitalist urbanization. Therefore, we

---

[1] The central government issued the Central Government's No.1 Document from 2004 to 2021 with the theme of "three rural areas" (agriculture, rural areas and farmers) for eighteen consecutive years.

introduce the concept of urban–rural integration in China and the empirical case of Hangzhou as a departing point of making contributions in this regard.

China's urban–rural relations in flux mark the paradigm shifts of China's urbanization [33]. In contrast to the urban–rural antagonism of capitalist development, socialist ideology in China points to a future of urban–rural integration [79]. Contemporary China sees rapid urbanization, but it comes with increasing urban–rural income disparity. According to [80], the per capita income ratio between urban and rural residents was 2.56 in 1978, which decreased to 1.86 in 1985 but spiked again to 3.3 by 2009. Additionally, the absolute gap widened from RMB4,027 to RMB12,021 during the same period. This striking fact indicates that income inequality has hindered the synchronization among urban and rural development, and income inequality is not the sole barrier to integrated urban–rural development at this stage. As [78] noted, four major issues confront the integrated urban–rural development strategy, namely: (1) income disparity, (2) waste of land due to scattered village layout, (3) inadequate public infrastructure and facilities, and (4) underdeveloped living environment and degraded natural environment. To address these issues, China has put consistent efforts into policy reforms. Since the formalization of coordinated urban–rural development at the 16th Party Congress in 2002, urban–rural integration has been prioritized in macro policy rhetoric. The development strategy of "rural revitalization" was proposed in the 19th Party Congress report in 2017, paying heightened attention to the rural areas. The most recent *Promotion Law of the People's Republic of China on Rural Revitalization*, released in April 2021, demonstrates the Chinese state's commitment to rural development. It is worthwhile to note that the ultimate goal of integrated urban–rural development is not to completely urbanize all rural territories or to homogenize both areas in terms of industrial structure and built environment, rather, to synchronize the development pace of both regions [81] and to create an equitable and harmonious society [82]. We will elaborate using the case of Hangzhou.

10.4.1 The Integrated Urban–Rural Development in Hangzhou

The 19th Party Congress Report highlighted the long-term plan for rural development, stating that a relevant institutional framework and policy system should be established by 2020, rural modernization being realized by 2035, and complete revitalization of the countryside be achieved by 2050. Against this backdrop, policy reform relating to urban revitalization has swept China. Being a representative of coastal metropolises and one of the pilot cities for policy initiatives in China [34], Hangzhou enjoys a relatively high urbanization rate at around 77 percent in 2019. Theoretically, Hangzhou has entered a stage of "fully urbanized" based on Kingsley Davis's S-curve [83], which lays a solid foundation for the promotion and implementation of urban–rural integration. Hangzhou is a sub-provincial city with a population of 10.36 million people that covers an area of 16,850 square kilometers. It comprises 10 districts, 1 county-level city, and 2 counties. Figure 10.1 shows the population change in the last couple of years in Hangzhou. From 2008 to 2019, the urbanization rate of Hangzhou increased from 69 to 77%, with a yearly increase rate of 0.7%. Meanwhile, the population of urban residents increased from 3.4 million to 5.36 million, with the proportion rising from 50 to 67%. The trend also indicates a significant decrease in the rural population. From 2008 to 2019, the rural population gradually decreased, with a plummet in 2014, but reached a plateau in [9]. The statistics showcase that the urban–rural population migration has entered a stable stage of two-way transition. While many rural residents seek better employment opportunities in cities, data show that 80% of the population still lives permanently in rural areas. In addition to the demographics, the employment structure in flux also suggests the trend of the declining primary sector and rising service sector (Fig. 10.2). Within the rural population, less than a quarter remains engaged in agriculture work. As [22] noted, the expansion of large-scale industrial agriculture has transformed the economic, environmental, and social structure of rural areas. While agriculture becomes industrialized and homogenized, the rural instead becomes much more diverse, dynamic, and vibrant with a gradual, tailored, and down-to-ground strategic development plan. Hangzhou has done a relatively successful job in balancing urban–rural development with the lowest income gap compared with other major metropolises in the Yangtze Delta Region and that of national. Indeed, very few cities have managed to keep this income ratio lower than 2, and Hangzhou has maintained the record since 2013 (Fig. 10.3). This accomplishment can be attributed to Hangzhou's long tradition of a rural economy [34] that siphons talents, capital, technology and other factors from urban. From 2009 to 2019, Hangzhou's urban per capita disposable income increased by 2.45 times, but that of rural areas increased by three times (Fig. 10.3). Although the absolute gap between urban (66,068 Yuan) and rural (36,255 Yuan) remains huge, the growth rate in rural areas is promising. All the statistics triumph are achieved through a development model other than urbanization that focuses on a one-directional channelling of resources, capital and people.

According to Zhejiang Urban and Rural Planning and Design Institute (ZURPDI), the two-way flow between urban and rural mainly attributed to three dynamics: (1) the maturity of the high-speed rail network and that mobile internet has weakened the attractiveness of the central city; (2) the urbanization has hit a bottleneck and stagnated in the eastern regions, rendering the two-way mobility trending; and (3) too high an urbanization rate (over 85% as in developed countries) is not suitable for traditional agricultural countries like China. This argument aligns with findings from existing literature, especially that the development of high-speed railway (HSR) is conducive to narrowing the income gap between urban and rural residents in China [85]. In addition, prior research also suggests that HSR reduces the health gap between urban and rural areas by enabling rural patients to seek medical treatments in cities [86]. The transport infrastructure, especially the logistics infrastructure in developed regions, has significantly and positively narrowed the urban–rural income gap [87]. The emerging e-commerce in Zhejiang province is also proved to have a significant role in narrowing the urban–rural income gap

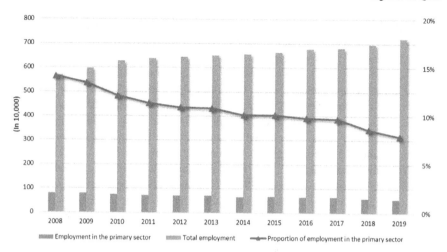

Fig. 10.2 Employment in the primary sector and total employment in Hangzhou, 2008–2019 (*source* [84])

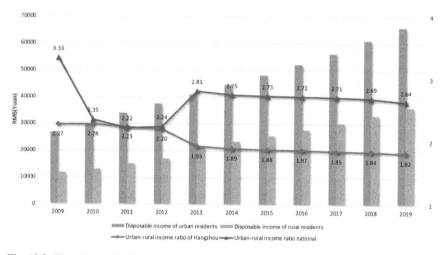

Fig. 10.3 The disposable income of urban and rural residents in Hangzhou, 2009–2019 (*source* [84])

[88]. Although HSR contributes to rural migration into cities, the migrant workers' improved mobility provides them with more flexibility in residential choices: rural residents can commute to work in cities [89]. Such dynamics remind us that traditional urban-oriented development strategies will give way to new integrated ones. In this context, the Hangzhou government has employed various strategies to facilitate the reorientation and integration of urban–rural development.

First off, it should be acknowledged that rural villages are heterogeneous regarding their historical, socio-economic, and geographical conditions. Therefore, the advancement of rural development should follow a strategy of classification as recommended by the *Rural Revitalization Strategic Plan (2018–2022)*. Four types of classified villages are identified: featured villages, clustered villages, urban peripheral villages, and deteriorated villages. Featured villages include famous historical and cultural villages, traditional villages, villages with minority characteristics, villages with featured tourism resources and others with rich natural historical and cultural resources. Clustered villages refer to the existing large-scale villages that are likely to survive in the future. Urban peripheral villages are suburban villages of central cities or county towns. The deteriorated villages are those to be demolished or consolidated for the suffering of natural hazards, fragile ecological environment, megaprojects development, and serious population loss. In 2018, there were 316 featured villages (13.2%), 1278 clustered villages (53.3%), 715 urban peripheral villages (29.8%), and 91 deteriorated villages (3.8%) in Hangzhou (Fig. 10.4). The structured classification of villages facilitates further planning and the implementation of tailored policies. For example, the development priorities for clustered villages include agricultural modernization, ecological conservation, and industrial upgrading (advocating for tourism and health industry and retirement services). While urban peripheral villages are more vulnerable to urbanization, especially urban expansion and annexation that fuels urban center development, other types of villages may not be subject to the extended urbanization process. Indeed, Hangzhou's rural planning is devised to reshape the centrality of rural, at least in the non-urban realm. As [71] argued, the urban–rural integration, "by rationalizing land use at a regional scale […], means that some areas are becoming more 'rural' as a result of urbanization" (p. 117).

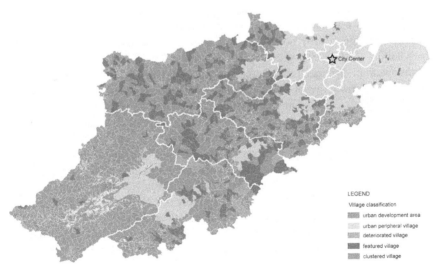

Fig. 10.4 The classification of villages in Hangzhou (*Source* reproduced by the author, Zhejiang Urban and Rural Planning and Design Institute.)

Second, industrial upgrading has been a potent tool in achieving urban–rural integration in Hangzhou. Urban agriculture, sustainable agriculture and digital agriculture have transformed and modernized traditional small- and medium-sized forms of food production. Prior research reported two primary economic modes that contribute to rural development: e-commerce based on the internet economy and the tourism economy [90–95]. The touristification in rural Hangzhou significantly contributes to the economic restructuring of the local areas. In 2017, around 56 million tourists visited rural Hangzhou, which accounts for one-third of the total tourists. Hangzhou government enacted targeted policies to support rural tourism, such as the talent attraction program. The attracted rural tourism entrepreneurial migrants not only bring new and creative development methods to the local areas but pitch the local tourism to regional audiences [96]. However, rural tourism in Hangzhou is also faced with challenges. According to a report by ZURPDI, the challenges include the lack of attractiveness, weak collective economies, the difficulty of rural land circulation, idle rural homestead lands, and ambiguous property rights. Such issues have led to Hangzhou's integrated way of industry upgrading.

Hangzhou initiated the "beautiful economy" that combines a diversity of economic modes in accordance with the *Hangzhou City Beautiful Countryside Construction Upgrade Action Plan (2016–2020)*. The "beautiful economy" includes but is not limited to tourism, health care, retirement housing and care services, Airbnb, e-commerce, sport and entertainment, and recreational parks. Among them, tourism is the most promising. According to statistics, tourism generated an income of 400 billion RMB in 2019, accounting for around 8% of Hangzhou's GDP [97]. Around 208 million tourists visited Hangzhou in 2019, of which rural tourism contributed 90 million. The municipal government of Hangzhou has deployed new development objectives for rural planning under the 14th Five-Year Plan, targeting a goal of 100 million tourists and 10 billion revenue income from tourism [98]. The economic diversity has significantly boosted rural residents' income and purchasing power. From 2000 to 2017, per capita disposable income increased by 6.2 times and per capita consumption expenditure increased by 6.5 times in rural Hangzhou, which also exceeded the growth rate of per capita income of urban residents in the same period [84]. The success is resultant of four strands of planning policy, focusing on: rural revitalization, landscape improvement, agricultural development, and performance management (Table 10.1). With the systematic planning guidance, local villages have undergone dramatic changes regarding industrial structure, lifestyle and rural environment. For instance, Daijia village have shifted from livestock farming in the 90 s to health and tourism industry from 2013. Currently, 85% of rural residents live in newly built townhouses (Fig. 10.5) and rely on "Nongjiale" (delights in farm guesthouses) as the primary source of income (30,000 to 40,000 RMB per year). To conclude, it is evident that the rural realm has played a proactive role in forging urban–rural integrated development under a top-down planning framework that leverages the rural resources to channel in capital and other urban resources.

Third, the innovative rural land policy reform has rendered the form of urban–rural integration more flexible. As we argued above, China's landownership dichotomy has fixed the land development model, making land resource allocation a one-way

Table 10.1 Performance management plan of Hangzhou agricultural office (2016–2020) (*source* translated from http://www.jxhz.gov.cn/szbmjxglgh/8326.jhtml)

Types	Project name	Objectives
Urban–rural coordination	Connecting villages	Establish a new round of municipal support group, raising funds to help achieve an average annual increase of 10 million yuan on the basis of 2016
	Increasing the income of low-income farmers	Increase the annual per capita disposable income of low-income farm households by more than 10% annually
	District, county, and municipality collaboration	Continuously enhance the comprehensive strength of the county economy and gradually narrow the gap between the development of counties (cities) and urban areas
Rural reform	Land titling and certificating	Complete the registration and certification of land rights within three years
Rural construction	Beautiful countryside	Creating about 50 fine model lines, about 30 style towns, about 20 provincial and municipal key villages of historical and cultural villages
	Rural garbage classification and resource-based treatment	Classify over 80% domestic waste, achieve a treatment rate of over 70% and the waste reduction rate of over 50%
	Creating "Hangzhou-style dwelling"	Complete around 20 Hangzhou-style dwellings by the end of October 2018
Rural E-commerce	City-wide rural e-commerce sales	By 2020, to achieve the city's rural e-commerce sales of 9 billion yuan
Rural development	Industrial park construction	Cultivate 30 industrial parks, achieving an output value of 600 million Yuan, and absorbing 1,500 rural laborers
	Accelerate the transformation and upgrading of farmhouses (nongjia le)	Entertaining more than 50 million tourists and raising income to 5 billion yuan
	New agriculture	The total number of family farms reached 300, and the total number of farmers' professional cooperatives reached 400

process. Through land expropriation, rural lands are accumulated for urban development use. Rural residents' lack of property rights over rural land hindered an effective circulation of land which can be mutually beneficial for both rural and urban areas. Notably, we are not referring to the farmland that underpins China's food security, but the idle rural construction land that has increased from 335.1 to 443.8 square meters per capital from 2010 to 2016. This land waste issue is partially due to the longstanding urban-biased development strategy. It has therefore expedited "the separation of three rights" concerning rural land [99, 100], namely the ownership,

Fig. 10.5 Rural landscape in Daijia village (*source* reproduced by the author, Zhejiang Urban and Rural Planning and Design Institute)

the contractual right, and the use right, which opens a window to alternative ways of rural–urban land transfer. The marketization of rural construction land renders the capital investment in rural areas possible. In this regard, two primary forms of development projects include rental housing and Nongjiale, which are officially advocated and endorsed by central China.[2] In this context, the concentration of rural population and the consolidation of rural land have created a unique rural landscape that does not resemble the "urban fabric" but of vernacular and rural characteristics.

The urban–rural integration is a gradual undertaking that is bound to confront difficulties. Major obstacles include predominant population outflow, weak collective economy, difficulties in land transfer in remote areas, construction land quota shortage, idle rural homestead land, and boundary conflict. Furthermore, the underdeveloped infrastructure, especially the lack of high-speed public transport connecting urban and rural areas and inadequate public facilities, has impeded the urban–rural integrated development. Even so, the case of Hangzhou has demonstrated that urbanization in China unfolds in a manner that departs from planetary urbanization, especially when it comes to the conceptualization and empirical evidence of urban–rural relations.

---

[2] The Pilot Program for the Use of Collective Construction Land for Rental Housing issued in 2017 and the Opinions on Supporting Entrepreneurship and Innovation of Returning Rural Workers to Promote the Integrated Development of Primary, Secondary, and Tertiary Industries in Rural Areas issued in 2016.

10.5 Concluding Remarks

As entering the era of urban, urban scholars are actively engaged in constructing a more inclusive and comprehensive conceptual framework for a better understanding of cities. The concept of planetary urbanization is an important contribution to urban geography in terms of the epistemological turn. This paper serves as more of a dialogue between the concept of planetary urbanization and the most contemporary urbanization practices in China rather than a critique of the theory. We argue that with the distinctive political-economic context, China's urbanization does not fully realize in the way portrayed by planetary urbanization. The integrated urban–rural development is a local approach at work in China, aiming to effectively balance development in both rural and urban areas. While planetary urbanization envisions the "complete urbanization of society," this is something that will not happen in China in the short term. On the contrary, China has paid remarkable attention to the rural side, prioritizing the revitalization of the countryside in its newly released 14th Five-Year Plan. As shown in the case of Hangzhou, urban–rural integration is the dominant model for urban–rural development in contemporary China. In this light, we believe it is critical to bring this discrepancy to the table to stimulate further discussions over the less attended topic.

The key to the explanation of this theoretical mismatch is somehow obvious. China has never embraced a capitalist mode of production completely, even with the proliferation of neoliberalism [101–103], and the authoritarian state are proactive in exercising its powerful planning intervention in urban development. Neoliberalization in China has strong 'Chinese Characteristics' [14, 102, 104] that is influenced by the "authoritarian turn" of neoliberalism [105]. So is the urbanization process in China. While planetary urbanization, especially the three moments—concentrated urbanization, extended urbanization and differential urbanization—best captures the capitalist mode of urbanization, it does not speak directly to the development of a country dominated by socialist ideology, even if it works in some ways. Although neoliberalism has penetrated into China's urban development, spawning many capitalist phenomena such as land commodification [106], land finance [107], and urban entrepreneurial government [108], the central government retains supreme authority over urban development planning. The decentralization of power and fiscal responsibilities in China has created a landscape of "urbanization of local state" [109], which ostensibly echoes [22], (p. 175) claim that "capitalist urbanization must be understood as a polymorphic, multiscalar and emergent dynamic of socio-spatial transformation." However, it is by no means to ignore the government's role in unpacking China's urbanization. The strong control over the direction of urban development underlies China's urbanization as an anomaly in planetary urbanization's prediction.

Planetary urbanization in China should thus be conceptualized with reference to the notion of urban–rural integration. Building on the three moments of urbanization (see Fig. 10.1 in [22]), we proposed a conceptual framework that constitutes four dialectic moments, which captures the ongoing urbanization process in China (Fig. 10.6). In contrast to the original framework, we highlight the role of rural

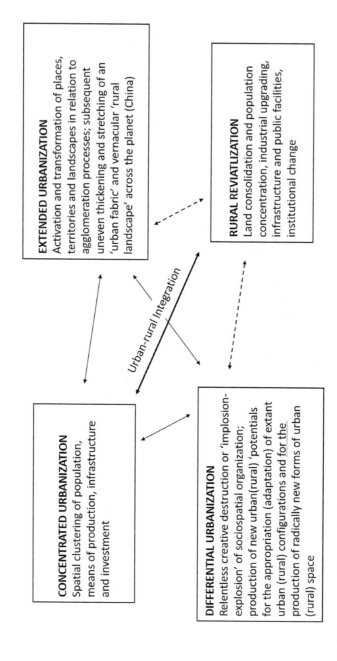

Fig. 10.6 A conceptual framework of "planetary urbanization in China"

revitalization as another important force that shapes the planetary landscape. Rural revitalization involves land consolidation and population concentration, industrial upgrading, the construction of infrastructure and public facilities, and institutional change. Such dynamics mutually influence both extended urbanization and differential urbanization. As such, the extended urbanization gives rise to not only the "thickening and stretching of an 'urban fabric'" but also a vernacular rural landscape across China. In the same vein, the differential urbanization alike produces new rural 'potentials' to adapt to extant rural configurations and to produce radically new forms of rural space. The urban–rural integration binds concentrated urbanization with rural revitalization through the planning intervention of China's state. While market force and capitalist mode of production maintain the interconversion of such moments, it is the authoritarian and blueprint planning that foregrounds the non-urban realm when unravelling China's contemporary urbanization. We believe that this framework is much pertinent to China's context and could inform further studies on China's urbanization.

It is worth noting that while the above framework speaks directly to China's urbanization process, it also reminds us of the ongoing endeavors of deciphering urbanization in the global south beyond the planetary urbanization paradigm and, broadly speaking, beyond the prevailing capitalist mode of production under neoliberalism [110]. Previous studies have offered ample empirical evidence in this regard. [111] contends that planetary urbanization generates 'novel socioenvironmental contradictions with a hybrid character of urban–rural-forest conditions' (p. 1083) in the Amazon region, and such relations warrant further inquiry on Latin American interpretations of new urbanity/rurality. Caldeira [112] argues that urbanization in the global south manifests in the form of peripheral urbanization that mainly deals with the production of residential urban spaces based on the cases of Istanbul, Santiago, and Sao Paulo. Rather than the capitalist logic of space production, Caldeira is concerned with residents' agency in producing urban spaces, which in his view is a process of *autoconstruction*. Reddy [113] intervenes with planetary urbanization from a postcolonial standpoint, maintaining that both urban and urbanization should be put under erasure to allow more possibilities of "context of contexts." Through a case study of the operational landscape of India's IT capital Bangalore, Reddy points out that "planetary urbanization fails to understand that global South cities [...] are simultaneously sites of concentrated, differential and extended urbanization" (p. 538). Lesutis [114] further notes that although the extractivism-induced urbanization in Tete, Mozambique, provides local residents with resettlement into urban space, the space does not function as urban. Specifically, residents living in such spaces are faced with unsafe and unstable living conditions which further displace them. Lesutis believes that the exact manifestation of planetary urbanization in the global south needs to be analyzed contextually. Furthermore, [115] caution us that the traditional perspective of urbanization in the global south often neglects the importance of demographic concentration in towns and small cities, which is a blind spot of planning policies as well as urban theories that invariably focus on megacities and agglomerations. Given the expansive and emergent body of knowledge regarding urbanization in the global south, it is therefore evident that our proposed framework has its innate constraints with its universality. However, it is our intention to offer observations of China's practices to foster dialogues among various countries of the Global South to advance mutual learning and development.

There are limitations to this paper. First, given the body of knowledge of planetary urbanization is expanding, this paper may not include timely readings and discussions, which may weaken the contributions made to the existing literature. Second, the integrated urban–rural development in China is heterogeneous due to the uneven geography of Chinese cities. The case of Hangzhou is representative of megacities in coastal regions but does not speak to cities in other regions, especially the medium- and small-size cities. Whether the proposed framework can be applied universally in explicating urbanization nationwide warrants further in-depth investigation and empirical studies. Third, the effectiveness of this development mode needs attentive evaluation. Recently, China has announced its complete victory in fighting against poverty. The urban–rural integration is integral to this success, but whether it can advance China's development one step further remains unknown. In addition, the question as to can this relative unique mode of development be translated and replicated in places of the Global South is yet to be answered.

Acknowledgements The authors would like to thank the anonymous reviewers for their constructive comments. We also thank Shanshan Ye for sharing the research reports from Zhejiang Urban and Rural Planning and Design Institute.

References

1. Brenner N, Schmid C (2014) The 'urban age' in question. Int J Urban Reg Res 38(3):731–755. https://doi.org/10.1111/1468-2427.12115
2. Lefebvre H (2003) The Urban Revolution. U of Minnesota Press
3. Harvey D (1989) The condition of postmodernity. Blackwell, Cambridge
4. de Leeuw S, Hunt S (2018) Unsettling decolonizing geographies. Geogr Compass 12(7):e12376. https://doi.org/10.1111/gec3.12376
5. Daigle M, Ramírez MM (2019) Decolonial geographies. In: Keywords in radical geography: antipode at 50. John Wiley & Sons, Ltd, pp 78–84. https://doi.org/10.1002/9781119558071.ch14
6. Sassen S (1991) The global city. New York
7. McGee TG (1991) The emergence of desakota regions in Asia: expanding a hypothesis. Settlement transition in Asia. University of Hawaii Press, The extended metropolis
8. Brenner N, Schmid C (2011) Planetary urbanization. In: Urban constellations. Berlin, Jovis, pp 10–13. http://urbantheorylab.net/site/assets/files/1016/2011_brenner_schmid.pdf
9. Derickson KD (2015) Urban geography I: Locating urban theory in the 'urban age'. Prog Hum Geog 39(5), 647–657. SAGE Publications Ltd. https://doi.org/10.1177/0309132514560961
10. Jazeel T (2018) Urban theory with an outside. Environ Plan D Soc Space 36(3), 405–419. SAGE Publications Ltd STM. https://doi.org/10.1177/0263775817707968
11. Callahan WA (2012) Sino-speak: Chinese exceptionalism and the politics of history. J Asian Stud 71(1), 33–55. Cambridge University Press

12. Ho B (2014) Understanding Chinese exceptionalism: China's rise, its goodness, and greatness. Alternatives 39(3), 164–176. SAGE Publications Inc. https://doi.org/10.1177/0304375414567978
13. Pow C-P (2011) China exceptionalism? Unbounding narratives on urban China. In: Urban theory beyond the West. Routledge
14. Harvey D (2005) A brief history of neoliberalism. Oxford University Press, USA, New York
15. Ma LJC (2002) Urban transformation in China, 1949–2000: a review and research agenda. Environ Plan A: Econ Space 34(9), 1545–1569. SAGE Publications Ltd. https://doi.org/10.1068/a34192
16. Haraway D (1988) Situated knowledges: the science question in feminism and the privilege of partial perspective. Fem Stud 14(3):575–599. https://doi.org/10.2307/3178066
17. Keil R (2017) Suburban planet: making the world urban from the outside in. John Wiley & Sons
18. Martinez R, Bunnell T, Acuto M (2020) Productive tensions? The "city" across geographies of planetary urbanization and the urban age. Urban Geography 1–12. Routledge. https://doi.org/10.1080/02723638.2020.1835128
19. Williams J, Robinson C, Bouzarovski S (2020) China's belt and road initiative and the emerging geographies of global urbanisation. Geogr J 186(1):128–140. https://doi.org/10.1111/geoj.12332
20. Wyly EK (2020) The new planetary suburban frontier. Urban Geogr 41(1), 27–35. Routledge. https://doi.org/10.1080/02723638.2018.1548829
21. Schindler S (2017) Towards a paradigm of Southern urbanism. City 21(1), 47–64. Routledge. https://doi.org/10.1080/13604813.2016.1263494
22. Brenner N, Schmid C (2015) Towards a new epistemology of the urban? City 19(2–3), 151–182. Routledge. https://doi.org/10.1080/13604813.2015.1014712
23. Schmid C (2018) Journeys through planetary urbanization: decentering perspectives on the urban. Environ Plan D Soc Space 36(3), 591–610. SAGE Publications Ltd STM. https://doi.org/10.1177/0263775818765476
24. Keil R (2020) After Suburbia: research and action in the suburban century. Urban Geogr 41(1), 1–20. Routledge. https://doi.org/10.1080/02723638.2018.1548828
25. Rogers S, Wilmsen B (2019) Towards a critical geography of resettlement. Prog Hum Geogr 030913251882465. https://doi.org/10.1177/0309132518824659
26. Roy A (2005) Urban informality: toward an epistemology of planning. J Am Plann Assoc 71(2):147–158. https://doi.org/10.1080/01944360508976689
27. Feldman S, Geisler C (2012) Land expropriation and displacement in Bangladesh. J Peas Stud 39(3–4), 971–993. Routledge. https://doi.org/10.1080/03066150.2012.661719
28. Sargeson S (2013) Violence as development: land expropriation and China's urbanization. J Peas Stud 40(6):1063–1085. https://doi.org/10.1080/03066150.2013.865603
29. Lipton M (1977) Why poor people stay poor: a study of urban bias in world development. Australian National University Press, Temple Smith
30. Oswin N (2018) Planetary urbanization: a view from outside. Environ Plan D Soc Space 36(3), 540–546. SAGE Publications Ltd STM. https://doi.org/10.1177/0263775816675963
31. Massey D (1991) Flexible sexism. Environ Plan D Soc Space 9(1), 31–57. SAGE Publications Ltd STM. https://doi.org/10.1068/d090031
32. Brenner N (2018) Debating planetary urbanization: for an engaged pluralism. Environ Plan D Soc Space 36(3), 570–590. . SAGE Publications Ltd STM. https://doi.org/10.1177/0263775818757510
33. Zhu J, Zhu M, Xiao Y (2019) Urbanization for rural development: spatial paradigm shifts toward inclusive urban-rural integrated development in China. J Rural Stud 71:94–103. https://doi.org/10.1016/j.jrurstud.2019.08.009
34. Qian Z (2015) Hangzhou. Cities 48, 42–54. https://doi.org/10.1016/j.cities.2015.06.004
35. Qian Z (2014) China's pre-reform urban transformation: the case of Hangzhou during the cultural revolution (1966–1976). Int Dev Plan Rev 36(2), 181–204. Liverpool University Press (UK). https://doi.org/10.3828/idpr.2014.12

36. Wang S, Tan S, Yang S et al (2019) Urban-biased land development policy and the urban-rural income gap: evidence from Hubei Province, China. Land Use Policy 87:104066. https://doi.org/10.1016/j.landusepol.2019.104066
37. Faas AJ, Jones EC, Tobin GA, et al (2015) Critical aspects of social networks in a resettlement setting. Dev Pract 25(2), 221–233. Routledge. https://doi.org/10.1080/09614524.2015.1000827
38. Price S (2019) Looking back on development and disaster-related displacement and resettlement, anticipating climate-related displacement in the Asia Pacific region. Asia Pac Viewp 60(2):191–204. https://doi.org/10.1111/apv.12224
39. Wang M, Lo K (2015) Displacement and resettlement with Chinese characteristics: an editorial introduction. Geogr Res For 35:1–9
40. Gao J, Liu Y, Chen J (2020) China's initiatives towards rural land system reform. Land Use Policy 94:104567. https://doi.org/10.1016/j.landusepol.2020.104567
41. Zhou Y, Li X, Liu Y (2020) Rural land system reforms in China: history, issues, measures and prospects. Land Use Policy 91:104330. https://doi.org/10.1016/j.landusepol.2019.104330
42. Chen JC, Zinda JA, Yeh ET (2017) Recasting the rural: state, society and environment in contemporary China. Geoforum 78:83–88
43. Su X, Qian Z (2020) state intervention in land supply and its impact on real estate investment in China: evidence from prefecture-level cities. Sustainability 12(3), 1019. Multidisciplinary Digital Publishing Institute. https://doi.org/10.3390/su12031019
44. Harvey D (1978) The urban process under capitalism: a framework for analysis. Int J Urban Reg Res 2(1–3):101–131. https://doi.org/10.1111/j.1468-2427.1978.tb00738.x
45. Fang C, Yu D (2017) Urban agglomeration: an evolving concept of an emerging phenomenon. Landsc Urban Plan 162:126–136. https://doi.org/10.1016/j.landurbplan.2017.02.014
46. Neef A, Singer J (2015) Development-induced displacement in Asia: conflicts, risks, and resilience. Dev Pract 25(5), 601–611. Routledge. https://doi.org/10.1080/09614524.2015.1052374
47. Tan Y (2020) Development-induced displacement and resettlement. Routledge, Routledge Handbook of Migration and Development
48. Salaff J (1967) The urban communes and anti-city experiment in communist China. China Quart (29). Cambridge University Press, School of Oriental and African Studies, pp 82–110
49. Cai F (2011) Hukou system reform and unification of rural–urban social welfare. Chin World Econ 19(3):33–48. https://doi.org/10.1111/j.1749-124X.2011.01241.x
50. Ma LJC, Fan M (1994) Urbanisation from below: the growth of towns in Jiangsu, China. Urban Stud 31(10), 1625–1645. SAGE Publications Ltd. https://doi.org/10.1080/00420989420081551
51. Liu YS (2019) Research on the geography of rural revitalization in the new era. Geogr Res 38(3):461–466
52. Yeh AG, Wu F (1999) The transformation of the urban planning system in China from a centrally-planned to transitional economy. Prog Plan 51(3):167–252. https://doi.org/10.1016/S0305-9006(98)00029-4
53. Chan KW (2010) Fundamentals of China's urbanization and policy*. China Rev Hong Kong 10(1), 63–93. Chinese University Press, Hong Kong, China, Hong Kong
54. Li L (2011) The incentive role of creating "cities" in China. China Econ Rev 22(1):172–181. https://doi.org/10.1016/j.chieco.2010.12.003
55. Liu Y, Yin G, Ma LJC (2012) Local state and administrative urbanization in post-reform China: a case study of Hebi City, Henan Province. Cities 29(2):107–117. https://doi.org/10.1016/j.cities.2011.08.003
56. Ma LJC (2005) Urban administrative restructuring, changing scale relations and local economic development in China. Polit Geogr 24(4):477–497. https://doi.org/10.1016/j.polgeo.2004.10.005
57. Liu Y, Li Z, Jin J (2014) Pseudo-urbanization or real urbanization? Urban China's mergence of administrative regions and its effects: a case study of Zhongshan City, Guangdong Province. China Review 14(1), 37–59. Chinese University Press, Hong Kong, China

58. Yew CP (2012) Pseudo-urbanization? Competitive government behavior and urban sprawl in China. J Contemp China 21(74):281–298. https://doi.org/10.1080/10670564.2012.635931
59. Lin GCS (2014) China's landed urbanization: neoliberalizing politics, land commodification, and municipal finance in the growth of metropolises. Environ Plan A Econ Space 46(8):1814–1835. https://doi.org/10.1068/a130016p
60. Andreas J, Zhan S (2016) Hukou and land: market reform and rural displacement in China. J Peasant Stud 43(4), 798–827. Routledge. https://doi.org/10.1080/03066150.2015.1078317
61. Huang D, Chan RCK (2018) On 'land finance' in urban China: theory and practice. Habitat Int 75:96–104. https://doi.org/10.1016/j.habitatint.2018.03.002
62. Ong LH (2014) State-led urbanization in china: skyscrapers, land revenue and "concentrated villages". China Quart 217, 162–179. Cambridge University Press. https://doi.org/10.1017/S0305741014000010
63. Zhang X, Li H (2020) The evolving process of the land urbanization bubble: evidence from Hangzhou, China. Cities 102:102724. https://doi.org/10.1016/j.cities.2020.102724
64. Wang X-R, Hui EC-M, Choguill C et al (2015) The new urbanization policy in China: which way forward? Habitat Int 47:279–284. https://doi.org/10.1016/j.habitatint.2015.02.001
65. Qian Z, Xue J (2017) Small town urbanization in Western China: villager resettlement and integration in Xi'an. Land Use Policy 68(Complete), 152–159. https://doi.org/10.1016/j.landusepol.2017.07.033
66. Yang C, Qian Z (2021) 'Resettlement with Chinese characteristics': the distinctive political-economic context, (in)voluntary urbanites, and three types of mismatch. Int J Urban Sustain Dev. https://doi.org/10.1080/19463138.2021.1955364
67. Lefebvre H (1991) The production of space. Oxford Blackwell, Oxford
68. Dolley J, Marshall F, Butcher B et al (2020) Analysing trade-offs and synergies between SDGs for urban development, food security and poverty alleviation in rapidly changing peri-urban areas: a tool to support inclusive urban planning. Sustain Sci 15(6):1601–1619. https://doi.org/10.1007/s11625-020-00802-0
69. Lo K, Wang M (2018) How voluntary is poverty alleviation resettlement in China? Habitat Int 73:34–42. https://doi.org/10.1016/j.habitatint.2018.01.002
70. Zhou Y, Guo Y, Liu Y, et al (2018) Targeted poverty alleviation and land policy innovation: Some practice and policy implications from China. In: Land use policy, vol 74. Land use and rural sustainability in China. pp 53–65. https://doi.org/10.1016/j.landusepol.2017.04.037
71. Wilczak J (2017) Making the countryside more like the countryside? Rural planning and metropolitan visions in post-quake Chengdu. Geoforum 78:110–118. https://doi.org/10.1016/j.geoforum.2016.10.007
72. Wu L, Zhang W (2018) Rural migrants' homeownership in Chinese urban destinations: do institutional arrangements still matter after Hukou reform? Cities 79:151–158. https://doi.org/10.1016/j.cities.2018.03.004
73. Cao Y, Bai Y, Zhang L (2020) The impact of farmland property rights security on the farmland investment in rural China. Land Use Policy 97:104736. https://doi.org/10.1016/j.landusepol.2020.104736
74. Cheng Y-S, Chung K-S (2018) Designing property rights over land in rural China. Econ J 128(615):2676–2710. https://doi.org/10.1111/ecoj.12552
75. Liu S, Carter MR, Yao Y (1998) Dimensions and diversity of property rights in rural China: dilemmas on the road to further reform. World Dev 26(10):1789–1806. https://doi.org/10.1016/S0305-750X(98)00088-6
76. Lu Q, Yao S (2018) From urban-rural division to urban-rural integration: a systematic cost explanation and Chengdu's experience. Chin World Econ 26(1):86–105. https://doi.org/10.1111/cwe.12230
77. Ye X, Christiansen F (2009) China's urban-rural integration policies. J Curr Chinese Aff 38(4), 117–143. SAGE Publications Ltd. https://doi.org/10.1177/186810260903800406
78. Zhang Z (2012) Theory, implementation and mechanism of urban-rural integration planning: a case study of Suzhou City. Mod Urban Res 4:15–20

79. Mili S (2019) Logical evolution of Marxist thought of urban-rural integration and development. MFSSR 2019:959–963
80. Sheng Z (2011) Towards China's urban-rural integration: issues and options. Int J China Stud 2(2), 345. Institute of China Studies, University of Malaya
81. Chen K, Long H, Liao L et al (2020) Land use transitions and urban-rural integrated development: theoretical framework and China's evidence. Land Use Policy 92:104465. https://doi.org/10.1016/j.landusepol.2020.104465
82. Chen C, LeGates R, Fang C (2019) From coordinated to integrated urban and rural development in China's megacity regions. J Urban Aff 41(2), 150–169. Routledge. https://doi.org/10.1080/07352166.2017.1413285
83. Davis K (1965) The urbanization of the human population. Sci Am 213(3), 40–53. JSTOR
84. Hangzhou Government (2020) Hangzhou statistical yearbook 2020. http://www.hangzhou.gov.cn/col/col805867/. Accessed 4 June 2021
85. Li W, Wang X, Hilmola O-P (2020) Does high-speed railway influence convergence of urban-rural income gap in China? Sustainability 12(10), 4236. Multidisciplinary Digital Publishing Institute. https://doi.org/10.3390/su12104236
86. Chen F, Hao X, Chen Z (2021) Can high-speed rail improve health and alleviate health inequality? Evidence from China. Transp Policy 114:266–279. https://doi.org/10.1016/j.tranpol.2021.10.007
87. Lu H, Zhao P, Hu H et al (2022) Transport infrastructure and urban-rural income disparity: a municipal-level analysis in China. J Transp Geogr 99:103292. https://doi.org/10.1016/j.jtrangeo.2022.103292
88. Li L, Zeng Y, Ye Z et al (2021a) E-commerce development and urban-rural income gap: evidence from Zhejiang Province, China. Pap Reg Sci 100(2):475–494. https://doi.org/10.1111/pirs.12571
89. Kong D, Liu L, Yang Z (2021) High-speed rails and rural-urban migrants' wages. Econ Model 94, 1030–1042. Elsevier
90. Gao J, Wu B (2017) Revitalizing traditional villages through rural tourism: a case study of Yuanjia Village, Shaanxi Province, China. Tour Manage 63:223–233. https://doi.org/10.1016/j.tourman.2017.04.003
91. Lin G, Xie X, Lv Z (2016) Taobao practices, everyday life and emerging hybrid rurality in contemporary China. J Rural Stud 47, 514–523. Rural Restructuring in China. https://doi.org/10.1016/j.jrurstud.2016.05.012
92. Peng C, Ma B, Zhang C (2021) Poverty alleviation through e-commerce: village involvement and demonstration policies in rural China. J Integr Agric 20(4):998–1011. https://doi.org/10.1016/S2095-3119(20)63422-0
93. Zhang H, Yan L-J, Zhang J-G, et al (2021) Rural landscape preferences and recreational activity inclination assessment from the tourist perspective, as linked to landscape values, in Deqing, China. Asia Pac J Tour Res 26(5), 488–503. Routledge. https://doi.org/10.1080/10941665.2020.1871386
94. Zhou J, Yu L, Choguill CL (2021) Co-evolution of technology and rural society: the blossoming of taobao villages in the information era, China. J Rural Stud 83:81–87. https://doi.org/10.1016/j.jrurstud.2021.02.022
95. Zou T, Huang (Sam) S, Ding P (2014) Toward a community-driven development model of rural tourism: the Chinese experience. Int J Tour Res 16(3), 261–271. https://doi.org/10.1002/jtr.1925
96. Li T, Li Q, Liu J (2021b) The spatial mobility of rural tourism workforce: a case study from the micro analytical perspective. Habitat Int 110:102322. https://doi.org/10.1016/j.habitatint.2021.102322
97. Hangzhou Culture, Radio, Film and Tourism Bureau (2020) 2019 Hangzhou tourism economic operation analysis. http://wgly.hangzhou.gov.cn/art/2020/1/27/art_1229278315_2539381.html. Accessed 30 May 2021
98. Hangzhou Government (2021) The next five years will usher in these changes in rural Hangzhou. http://www.hangzhou.gov.cn/art/2021/1/29/art_812262_59026572.html. Accessed 30 May 2021

99. Dong X (2019) The circulation of housing land use right in the past 70 years since the founding of new China: institutional change, current dilemma and reform direction. Chinese Rural Econ 6:2–27
100. Wang Q, Zhang X (2017) Three rights separation: China's proposed rural land rights reform and four types of local trials. Land Use Policy 63:111–121. https://doi.org/10.1016/j.landusepol.2017.01.027
101. Horesh N, Lim KF (2017) China: an East Asian alternative to neoliberalism? Pac Rev 30(4):425–442. https://doi.org/10.1080/09512748.2016.1264459
102. Ong A (2007) Neoliberalism as a mobile technology. Trans Inst Br Geogr 32(1):3–8
103. Zhou Y, Lin GC, Zhang J (2019) Urban China through the lens of neoliberalism: is a conceptual twist enough? Urban Stud 56(1):33–43. https://doi.org/10.1177/0042098018775367
104. Peck J, Zhang J (2013) A variety of capitalism with Chinese characteristics? J Econ Geogr 13(3):357–396. https://doi.org/10.1093/jeg/lbs058
105. Peck J, Theodore N (2019) Still neoliberalism? South Atlant Quart 118(2):245–265. https://doi.org/10.1215/00382876-7381122
106. Liu Y, Yue W, Fan P et al (2016) Financing China's suburbanization: capital accumulation through suburban land development in Hangzhou. Int J Urban Reg Res 40(6):1112–1133. https://doi.org/10.1111/1468-2427.12454
107. Qun W, Yongle L, Siqi Y (2015) The incentives of China's urban land finance. Land Use Policy 42:432–442. https://doi.org/10.1016/j.landusepol.2014.08.015
108. Chien S-S (2013) New local state power through administrative restructuring—a case study of post-Mao China county-level urban entrepreneurialism in Kunshan. Geoforum 46:103–112. https://doi.org/10.1016/j.geoforum.2012.12.015
109. Hsing Y-T (2010) The great urban transformation: politics of land and property in China. Oxford University Press Cary, Oxford
110. Parnell S, Robinson J (2012) (Re)theorizing cities from the global south: looking beyond Neoliberalism. Urban Geogr 33(4):593–617. https://doi.org/10.2747/0272-3638.33.4.593
111. Kanai JM (2014) On the peripheries of planetary urbanization: globalizing manaus and its expanding impact. Environ Plan D Soc Space 32(6), 1071–1087. SAGE Publications Ltd STM. https://doi.org/10.1068/d13128p
112. Caldeira TP (2017) Peripheral urbanization: Autoconstruction, transversal logics, and politics in cities of the global south. Environ Plan D Soc Space 35(1), 3–20. SAGE Publications Ltd STM. https://doi.org/10.1177/0263775816658479
113. Reddy RN (2018) The urban under erasure: towards a postcolonial critique of planetary urbanization. Environ Plan D Soc Space 36(3), 529–539. SAGE Publications Ltd STM. https://doi.org/10.1177/0263775817744220
114. Lesutis G (2020) Planetary urbanization and the "right against the urbicidal city". Urban Geogr 1–19. Routledge. https://doi.org/10.1080/02723638.2020.1765632
115. Randolph GF, Deuskar C (2020) Urbanization beyond the metropolis: planning for a large number of small places in the global south. J Plan Educ Res. SAGE Publications Inc. 0739456X20971705. https://doi.org/10.1177/0739456X20971705

Chapter 11
The Transport System as a Process of Territorial Development or the Risk of Social Segregation

Riad Arrach, Amal El Jirari, and Oussama Benmakrane

11.1 Introduction

Similar to socio-economic changes and demographic transformations, urban fabrics are now undergoing irrevocable socio-spatial recomposition. Peri-urban spaces are essential in the sense that they ensure the transition from rural to urban space. This further absorbs the pressure exerted on major polarities. It is true that peri-urbanization asserts itself as an effective solution to accommodate for the challenges of demographic growth. However, this solution is replete with several issues relating to mobility, housing, employability and living environment.

These issues represent a great responsibility in the hands of the leaders and officials who pull the strings of spatial organization and mobility in the various departments of the region. Poor management of this process can lead to serious repercussions and destabilize the proper functioning of society. Ensuring urban development and the smooth running of an urban transport system is a delicate but essential task for any society.

Therefore, intriguing questions arise, who controls the development process at the periphery or the latter is it totally out of control? How can the public transportation system be poorly organized? does these failures lies at the administrative and governance level or at the operational level? Does the Rabat-Témara conurbation experience good management in its development process and its collective transport

---

R. Arrach (✉) · A. El Jirari · O. Benmakrane
National Institute of Planning and Urbanism, Rabat, Morocco
e-mail: riadarrach@hotmail.fr

A. El Jirari
e-mail: amal.eljirari@hotmail.fr

O. Benmakrane
e-mail: o.benmakrane@gmail.com

© The Author(s), under exclusive license to Springer Nature Singapore Pte Ltd. 2023
L. Zhang et al. (eds.), *The City in an Era of Cascading Risks*, City Development: Issues and Best Practices, https://doi.org/10.1007/978-981-99-2050-1_11

system or is it bathed in disorder? We will try through this article to answer these questions.

The Rabat-Témara conurbation has undergone profound spatial changes and major socio-economic and demographic upheavals leading to the emergence of heterogeneous peri-urban spaces. The peripheral area of the Rabat-Témara agglomeration is now facing several issues and challenges relating to accessibility, functional positioning and the quality of the urban landscape.

Indeed, demographic growth and uncontrolled urban expansion have generated an inequitable distribution of resources, a lack of equipment and infrastructure and a fragmentation of the peri-urban fabric. Furthermore, the peri-urban area of the Rabat-Témara conurbation suffers from the poor operationality of the transport system, generating disarticulation between the different territorial entities and causing the aggravation of the segregation phenomenon.

This paper attempts to test the hypothesis that peri-urbanization causes social segregation, which is accentuated by a failing transport system. The objective of this analysis is to reflect upon both the phenomenon of peri-urbanization and the resulting socio-spatial segregation through the analysis of the transport system connecting the conurbation of Rabat-Temara to its hinterland.

The study area includes the conurbation of Rabat-Témara and its peripheral zone consisting of four peripheral municipalities namely; Ain Atik, Mers El Khir, the new town of Tamesna and El Menzeh (Fig. 11.1). The study area is bounded by Oued Bouregreg to the North, by the Atlantic Ocean to the East, by the plateaus of Shoul and Sidi Yahya Zaire to the West and by the Casablanca-Settat region to the South. According to the High Commission for Planning, the study area covers an overall area of 188,600 Ha and includes more than 998,814 inhabitants in 2014.

11.2 Literature Review

11.2.1 Peri-Urbanization: Conceptual Framework and Explanatory Factors

Peri-Urbanization: Diversity of Definitions

According to [4], "the phenomenon of urban transition is characterized by the transition from the homogenous function of the rural settlements and spatially dispersed villages to a more heterogeneous settlement concentrated in and around urban centers".[1]

Peri-urbanization is generally understood by an overlap between the characteristics of the urban environment and the specificities of the rural fabric. This results

---

[1] Bogaert and Halleux [4] Territoires périurbains: développement, enjeux et perspectives dans les pays du Sud. Belgique: Presses agronomiques de Gembloux.

11 The Transport System as a Process of Territorial Development …

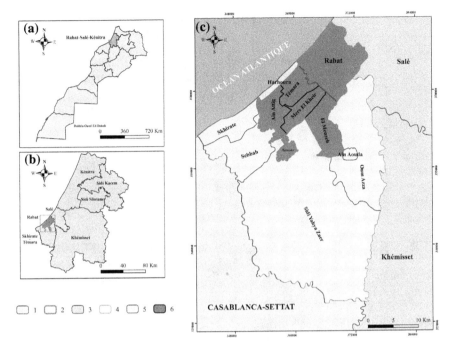

Fig. 11.1 Situation map, **a**: the study area in the national context, **b**: the study area in the regional context, **c**: the study area in the local context, 1: territorial organization of morocco, 2: limits of the region of rabat-salé-kénitra, 3: provincial division of the region of rabat-salé-kénitra, 4: limits of the prefecture of skhirate-témara, 5: municipal division of the prefecture of rabat and skhirate-témara and 6: study area

in a conceptual overlap where the emergence of multiple definitions impedes a terminological consensus.

In addition, the debate around peri-urbanization brought into the surface questions concerning the delimitation of these peri-urban areas. An interesting number of academics argue that the peri-urban perimeter begins just after the urban ring. Other researchers, on the other side of the coin, consider that the peri-urban comprises all of the rural hinterland marking the beginning of an eventual urbanization.

The National Institute of Statistics and Economic Studies suggests that the delimitation of urban areas is subject to two main criteria: first, the criterion of spatial morphology with respect to the discontinuity of the building and second, the dependency of peri-urban areas on employment availabilities in metropolitan France.

In order to deal with the conceptual ambiguity of the notion of peri-urbanization, it would be interesting to propose a definition by identifying the specificities of the peri-urban fabric. In this vein, several characteristics are to be emitted, Bruck[2] (2006, p.59):

- Peri-urban fabrics are generally characterized by a pervasive landscape combining aspects of both the rural and urban environment, thus generating a lack of territorial identity. Also, rapid urban expansion tends to hinder the identification of peri-urban boundaries, which further complicates our understanding of this phenomenon.
- The weak connection of the peri-urban setting with its metropolitan environment generates a strong dependency relationship between the polarity and its peripheries.
- Peri-urban spaces maintain relationships of interdependence and functionality with centrality in terms of access to structuring facilities and to the employment area.

One may conclude, in this regard, that the definition generated through the enumeration of the different characteristics of the peri- urban area shows that this phenomenon is much more complex and therefore difficult to dissect. The peri-urbanization process is not universal, but rather differs from one context to another and from one territory to another. Peri-urbanization can thus take various criteria and diverse forms,[3] [9] it presents different challenges that prevents the phenomenon to be identified in a unanimous and consensual manner.

Peri-Urbanization: A Multi-factor Phenomenon

In order to better understand the phenomenon of peri-urbanization, it would be interesting to analyze the underlying causes that led to the current state of events. Several factors are to be discussed:

- According to several authors, demographic growth is the main factor responsible for the phenomenon of peri-urbanization. The socio- demographic transformation has been frequently accompanied by a natural increase in housing demand. This, in turn, has accentuated the emergence of peri-urban spaces.
- In light of the above-mentioned factor, the housing supply cannot be met in the center given the financial incapacity of peri-urban families to access property situated in urban areas.
- Peri-urban areas are characterized by a relatively abundant land supply compared to the central city, allowing rapid urban expansion and meeting housing needs at an affordable cost.

---

[2] Bruck, L. (2006) La périurbanisation en Belgique: comprendre le processus de l'étalement urbain. GEO Géographie. Écologie-environnement. Organisation de l'espace, p.59.

[3] Ghezal and Bouchemal [9] Durabilité et périurbanisation: cas de la ville de Hamma Bouzian. Mémoire pour l'obtention du diplôme de Master en urbanisme. Université d'Oum El Bouaghi. Algérie.

- Beyond the availability of land which favors urban expansion, the tax advantages attract more investment, creating a peri-urban dynamic.
- The development of peri-urban areas is highly dependent on the development of a transport system connecting the peripheral areas to the center, a factor that expand the phenomenon of peri-urbanization.

11.2.2 A Failing Transport System: From Peri-Urbanization to the Accentuation of Socio-Spatial Segregation

Peri-urbanization is a complicated process of urban expansion. In their book "Peri-urbanization: issues and perspectives",[4] the authors denounce the complex nature of the phenomenon of peri-urbanization. The latter is traditionally appreciated as a process of transition to urbanity. However, Roux, Vanier [17] argue against and state the way in which peri-urbanization creates inequalities between territories and segregations between different social classes.

In fact, peri-urbanization is a double-edged phenomenon guaranteeing access to the resources and advantages of the metropolis while condemning the peri-urban population to live in relatively worrying conditions. Based on this premise, the gaps between the metropolis and the peripheries tend to widen. Moreover, the article by [1],[5] stipulates that the phenomena of peri-urbanization amplify the generating mechanisms of social segregation. The author emphasizes how the cumulative difficulties of mobility are the main intensifiers of the inequalities between the center and the periphery.

Charlot [7][6] affirmed that the process of peri-urbanization is accompanied by a phenomenon of social segregation within the peripheral zones insofar as it exerts a selective sorting between the populations. The Rabat-Témara agglomeration reflects a heterogeneous hinterland characterized by deep peri-urban areas with high urban density. The socio-spatial segregation between the Rabat-Témara agglomeration and the peri-urban municipalities is observed through several levels of analysis:

- On the demographic level, the population is unevenly distributed over the territory, favoring an anarchic development of peri-urban areas, which leads to the formation of imbalances between the centrality and its peripheries.
- The agglomeration is experiencing an unequal distribution of the supply of public transport. The presence of a green belt in the peripheral areas of the Rabat-Témara

---

[4] Roux and Vanier [17] La périurbanisation: problématiques et perspectives. La Documentation française. DIACT. p. 87.

[5] Ariotti [1] Les effets de la périurbanisation amplifient les inégalités sociales. Les Cahiers du Développement Social Urbain À La Voulte-sur-Rhône. 1(1), 14-14. https://doi.org/10.3917/cdsu.053.0014.

[6] Charlot et al [7] Périurbanisation, ségrégation spatiale et accès aux services publics. Ministère de l'Equipement, des transports, de l'Aménagement du territoire, du Tourisme et de la Mer.

agglomeration impedes the development of a transport system and road infrastructure that would ensure the connection between Rabat and the neighboring peri-urban municipalities.
- The metropolitan character of the Rabat-Témara agglomeration favors the development of hybrid peri-urban areas with alarming poverty rates and fragmented urban fabrics.
- The urban sprawl of the Rabat-Témara agglomeration results in social stratification. The poorest social classes tend to concentrate in peri-urban areas due to the accessibility of housing.

11.3 Methodology

This research follows the hypothetico-deductive approach through which I intend to develop a double hypothesis strongly imbued with the theories discussed and analyzed at the end of the literature review. The formulation of this hypothesis is the fruit of a confrontation between paradigms and doctrines that have dealt with issues relating to peri-urbanization, social segregation and urban mobility. These theoretical hypotheses will be the subject of an empirical test through a case study of the Rabat-Témara agglomeration for a possible confirmation or invalidation.

This paper is based on a qualitative analysis. The latter includes collected data from different sources on public transport and social segregation in peripheral areas and through semi-directive interviews with managers and some users of these means of transportation. The method adopted with the users is based on the random selection of samples. A review of the literature (articles, dissertations, books, research work, theses) and analysis of the qualitative data were also carried out.

Finally, in order to better contain the phenomenon of peri-urbanization and to understand the transport system, we used S.I.G tools under the ArcGIS 10.4 program as a methodology for mapping and spatial analysis. This would allow us to develop a spatially referenced database for all stages of the research. The geographic information system not only makes it possible to organize, restore and process databases, but also to map phenomena that are difficult to define.

Figure 11.2 shows the three steps followed in our methodological approach.

11.4 Results and Discussions

11.4.1 The Role of Public Transport in Urban Space

In order for a city to maintain a link between its territorial entities, it must go through its transport networks and the means that contributes to it, namely the transportation and travel system. In fact, a transport system that consists of roads and locomotion modes as infrastructure (bus, taxi …), should provide easy links to every city level.

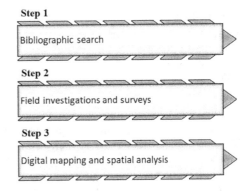

Fig. 11.2 The steps followed for the research methodology

Therefore, urban transport is of vital and strategic importance. It shapes the availability of individuals, goods and services at the right time and in the right place, and carrying out everyday life activities[7] [18].

Transportation (road infrastructure, railways, etc.) and mobility (circulation of individual vehicles, management of collective transport) are implemented to meet the need of moving from a place to another. Transport, through its organization and operation, helps define the spatial structure of the city.

The transformation of spatial structures by extension (development along axis or creation of peripheral housing poles), or by densification of the urban area, modifies the transport supply as well as the volume and distribution of travel demand[8] [12], p.4). Changes in cities appear to have implications for travel and transport supply, which in turn affect how cities evolve.

There appears to be a close interdependence between urban spatial growth and transport. It is therefore essential to combine urban planning and travel into the same urban planning process.

11.4.2 The Relationship Between Transport Urbanization

The "transport-urbanization" relationship refers to the system that characterizes the city. In fact, the city is a complex system which analysis requires a systemic approach that emphasizes the relationships between the different components of the urban system. We must first detach ourselves from the precarious idea that the city is essentially merely based on the simple aspect of the urban landscape, the use of ground, or on the description of demographic characteristics and economic activities

---

[7] Zahiri [18] Urbanisation, déplacements et transports urbain dans le grand Agadir. Mémoire de 3ème cycle pour l'obtention des études supérieures en aménagement et urbanisme. p.6.

[8] J Masson [12] Les interactions entre système de transport et système de localisation en milieu urbain et leur modélisation. Thèse de doctorat en sciences économiques, Mention Economie des transports. France: Lyon 2, p.4.

or even the classification of the standards of living. It is rather the combination of all these multiple characteristics and the complexity of their interrelations[9] [3].

It is also important to understand that the functioning of the various sub-systems that make up the urban space and particularly transport and locations of households and activities, results in a more complex interaction. It is the behavior of the population and the economic actors with respect to travel demand and residential rent and employment that determine the functioning of an urban space.

The last century has witnessed a considerable evolution in the transport system. This has been characterized by a growth in daily mobility (mainly focusing on car mobility), the decline of public transport, a high increase in long-distance travel and a geographical fragmentation of trips.

This evolutionary trend has come with two aspects. First, a strong development of rural areas, part of the industrial settlement, and also in urban areas by pushing the cities to expand over the nearby peripheries, also known as the peri-urbanization phenomenon. Second, a development of the vast metropolitan areas characterized by an imbalance between housing and employment.

According to Poulit[10] (1997), transport policies are the key to a good urban policy. In fact, the location of transport stations, the service of railways line with several terminus, for instance, must be deployed according to the objectives of town planning. Similarly, the choice of priority infrastructure projects must ensure directing urbanization in accordance with the urban policy pursued.

Also, public transport pricing policy can be an effective tool for guiding the process of urban growth. Lower rates in conjunction with long distance leads to a reduction in the density and a decrease in the number of people in the center zone. Higher rates, however, can lead to a concentration of urbanization and human settlement.

Since gaining their independence, the modern metropolises of developing countries such as the countries in North Africa have recorded the highest demographic growth in the world.

The road network has been the main support for this increased urbanization. Guided by the need for employment and transport facilities, the country's largest population tend to settle in the major urban centers. This extreme polarization of urban space has aggravated road traffic problems.

Moreover, urban centralization presents a number of drawbacks, such as road traffic, rise in land and rent prices, and abundance of undeveloped lots farther from the center, which explains the peri-urbanization phenomenon. This is a worrying situation given the inadequate spatial management of urban activities results in dispersed urban forms of the city, often difficult to serve by public transport[11] and therefore inaccessible for population who cannot afford to own car.

---

[9] Beaujeu-Garnier [3] Géographie urbaine. Armand Colin, p.25.

[10] Poulit, J. (1997). Les enjeux économiques et environnementaux de la mobilité, la jaune et la rouge, France.

[11] CETUR [6] Les enjeux des politiques de déplacement dans une stratégie urbaine. P. 46.

Public policy of the road network has not always preceded or accompanied urban development.[12] Public transport network is generally the result of a historical development which has not responded to the accelerated growth or the structural change of urban areas. This has led not only to a lack of integration of certain areas, but also to an insufficient coverage of the territory.

Morocco, like many developing countries, witnessed after its independence an increasing level of demographic growth and an accelerated tertiary. The new state of events was generally due to the massive rural exodus, a relative diversification of the urban economy as well as the transformation of social relations. The speed and the scale of this urbanization process made it difficult to manage this influx of people into the city. The progression of urban sprawl and the ongoing development of peripheral districts could, if nothing is considered to limit these phenomena, give rise to a metropolitan urban structure unsuitable for collective mobility.

The city of Rabat and Temara, like a good number of large Moroccan cities, are experiencing sustained dynamics at all levels: significant demographic growth, developing tourist infrastructure, emerging industry, rampant urbanization, etc. This has given rise to significant needs for exchanges and communication and has generated an increasingly important mobility of population in quest of employment and services.

With that being said, planning actions for the development of the conurbation and its surroundings have been carried out over the past 25 years. Notably, the 1991 SDAU seeks to curb property speculation and improve circulation and public transportation.

According to a critical evaluation of the implementation of the SDAU guidelines, an "urban project for the agglomeration of Rabat" was presented by the Rabat-Salé Urban Agency in 2004. This policy document established a framework for urban actions making it possible to strengthen the attractiveness of the city, to raise the urban quality and to connect the conurbation to its periphery.[13] One of the five major scopes of the Rabat urban project deals with the construction of metropolitan infrastructures, in particular the creation of a tram line and a motorway bypass.

Morocco is also engaged in a preparation of a national strategy for urban transport (2008) where the improvement of the conditions of public transport in each agglomeration constitutes one of the main axes around which this strategy must be implemented.

In fact, the lack of a comprehensive vision for urban planning has led to a dysfunction of the urban framework in the conurbation of Rabat-Témara. A dysfunction characterized by inequalities in the distribution of equipment, infrastructure and jobs, and an anarchy in the organization of transport. Morocco through its new strategy aims to repair public transport from these dysfunctions within the framework of a new global urban transport policy.

---

[12] Certu [5] Mobilités et transports, Fiche n°1, Comment élaborer des stratégies de mobilité durable dans les villes des pays en développement ? Note de synthèse p.8.

[13] Ministère de l'urbanisme et de l'aménagement du territoire [14] SDAU 2040 Rabat-Salé-Témara et sa zone périphérique. Maroc: Direction de l'urbanisme, p.17.

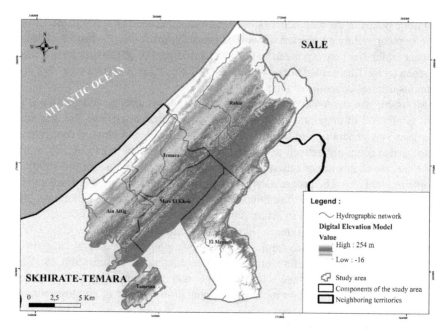

Fig. 11.3 Cartographic interpretation of the physical situation of our study area

These measures were fulfilled. However, the accessibility of the peripheral zones to the urban center remains a big issue until now. This represents a key indicator on which we must base our focus in order to reconcile urban planning and transport policies. Dealing with only one at a time and not both simultaneously will always lead us to social, economic and demographic imbalances.

Conurbation of Rabat Temara and its periphery currently has a number of inhabitants exceeding a million. The fluctuation of its population growth is characterized by a slight decrease at the city of Rabat, the growth rate rose to -1.29% between 2004[14] and 2014.[15] On the other hand, the municipalities of peripheral areas growth rate experienced a demographic explosion: a rate of 42.30% in the municipality of Mers El Kheir, 70.49% in Ain Attiq and 89.53% in the municipality of El Menzeh between 2004 and 2014. Moreover, population concentration of the new city of Tamesna exceeds by far all the other common peripheral (see Fig. 11.5).

Before developing the analysis on the data that we have just mentioned, it is essential to understand the real physical and social situation of our field of study, this will be the subject of a cartographic interpretation concerning the physical situation of our conurbation (Fig. 11.3) and a graphical interpretation of the social situation

---

[14] Haut-Commissariat au Plan [10]. Recensement générale de la population et de l'habitat de 2004. Maroc.

[15] Haut-Commissariat au Plan [11]. Recensement générale de la population et de l'habitat de 2014. Maroc.

11 The Transport System as a Process of Territorial Development ...

(Fig. 11.4). The choice of interpretation mode is related to the nature of the data used.

The following graph analyzes three important data concerning the social situation, namely demography, illiteracy and the level of study, we note that the conurbation

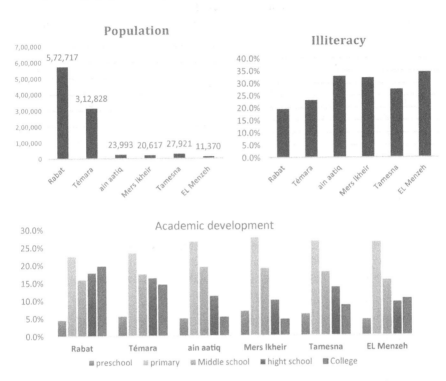

Fig. 11.4 Graphic representation of the social situation of rabat temara and its outskirts

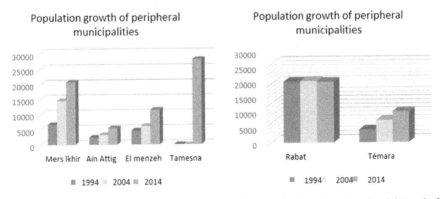

Fig. 11.5 The demographic distribution of the conurbation of rabat-témara (on the right) and of the peripheral municipalities (on the left)

has the lowest illiteracy rate unlike the periphery. For the level of study, we note that the rate of the population having studied at college is three times higher than that living in the periphery.

Regarding income data, unfortunately it is not published at the level of the High Commission for Planning or even at the level of the Ministry of Finance, it remains confidential. If we had access to this data, the analysis would be more relevant. Figure 11.4 shows graphic interpretation of the social situation of the conurabation.

The polarity map (Fig. 11.6) of the study area gives an idea of the demographic concentration at the level of the different municipalities.

This demographic shift is due to the fact that much of the demographic weight of the conurbation migrates to the hinterland, a periphery of Rabat-Témara. The peripheral municipalities act as a sponge that absorbs the demographic surplus experienced by the conurbation.

It remains to be seen if this demographic and urban development is accompanied by an adequate transport system that effectively connects the conurbation to its peripheral areas.

The following map shows the distribution of bus lines between the conurbation and its peripheral area (Fig. 11.7).

The distribution of the bus lines is as follows:

Of the 17 bus lines available, four lines go through Ain Attiq, two lines through Mers El Kheir and Tamesna and one line go through El Menzeh. We notice that the service is poorly organized spatially, in addition to this, according to several

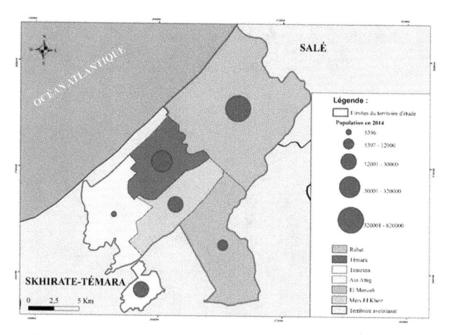

Fig. 11.6 Polarity map of the study area

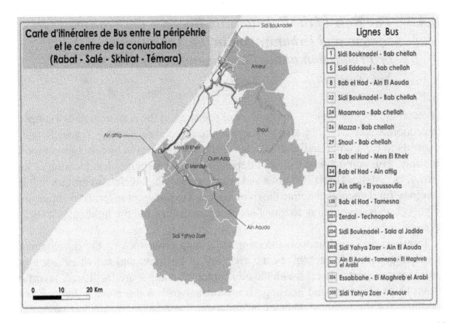

Fig. 11.7 Route map of bus lines the conurbation and its peripheral area (Nduwima and al 2020)[16]

testimonies from bus users, the frequency is not respected and the price is deemed too high.[17] What's more, the public transport supply is not properly adapted to the demand of the populations living in the peri-urban circle. In fact, the development of urban planning and the supply of public transport offer in the case understudy are at a different pace of development.

Despite the economic and social importance of public transport in the daily life of citizens and the plurality of public and private actors operating in the public transport sector for decades, we are witnessing a rupture between urban planning and public transportation. The latter can be at the service of the urban project through channeling planned urbanization and influencing household residential choices. It seems that public transport is highly important in public policies for a land planning and mobility management. To this very day, Rabat and Témara do not have an urban travel plan. In this way, this paper will greatly contribute to the management of mobility.

---

[16] Nduwima [15] Emploi, transport et mobilité. Rapport de diagnostic. Maroc: INAU.
[17] Nduwima [15] OP. Cit, p. 21.

11.4.3 The Role of the Relationship Between the Transport System and Urbanization in the Accentuation or Production of the Phenomenon of Social Segregation

It is essential to underline the importance of transport in the conurbation as a support for economic activity. They also play a prominent social role as they connect neighborhoods to each other and to the center of the agglomeration or conurbation more precisely. This role may be different depending on the geographical location of the district, its distance from downtown and urban services and its links to the city.[18] It is, therefore, fundamental to define the roles assigned to transport in terms of enhancing neighborhoods, responding to travel needs, accessibility and the fight against social segregation.

Transport and road networks determine, among other factors, the development of neighborhoods. In fact, given the diversity of business and activities' location linked by transport networks with the city center, exchanges are facilitated, favoring access to employment and housing. Transport network is therefore a sine qua non condition for the establishment of professional, personal or leisure activities and can help reduce the mono-functionality of certain districts. However, this valuation requires a favorable environment that is more accessible to neighborhoods located near urban centers and to economically dynamic employment areas.

The development of a district can also be accompanied by a redistribution of its populations, the arrival of more affluent social strata and the aggravation of social segregation in the most distant places.

Merlin [13], p. 46)[19] specifies that *"the purpose of urban transport is to provide city dwellers with accessibility to other elements of the city. In developed countries, the development of individual means of transport and the ramification of public transport networks seem to make this reminder superfluous. This is not quite correct: think of people in a small town without public transport; people living on the fringes (disadvantaged neighborhoods) or sometimes beyond the agglomeration and working in cities. This is even less true in the case of cities in developing countries where often entire neighborhoods, spontaneously developed on the periphery, are not served. Access to the city, to a job, requires, for those who are not the happy owners of a car or a bicycle, hours of walking. Accessibility then becomes an absolute priority [...] The major objective of any urban transport policy must therefore be to ensure, for the entire population, the possibility of accessing employment areas and downtown facilities"*.

---

[18] CETUR [6] Op, Cit., p. 47.

[19] Merlin [13] La Planification des transports urbains. Paris: Masson et Cie, p.46.

Equal opportunity and the accessibility to all activities (employment, gym, entertainment, culture, services, shops) is what constitutes the right to transport.[20] Accessibility presupposes first to remove obstacles of all kinds, like physical ones, through the establishment of a proper road network with public transportation means and the creation of infrastructure that facilitate non-motorized travel like walking and cycling [2]. Accessibility would also mean removing all form of informational and pricing constraints.

The pricing policy must in fact be adapted to the wages of the most vulnerable groups and households. This requires a detailed knowledge and awareness of the needs of this social class. This would make it possible to define a suitable price for each locality and that is adapted to the average purchasing power of its inhabitants.

Accessibility is a vector of social cohesion. With the increase of urban rhythms, the gaps between populations are likely to increase and inequalities follows the same path with mobility practices, immobility or "reduced mobility" becoming a factor of increased segregation for various categories of people. It is through accessibility that it is possible to consider people who are less mobile, by choice or by constraint and who, nevertheless, want to access the activities of the city.[21]

According to this logic, transport play a prominent role insofar as it is situated "on the territory" and connects the latter by the contribution of the deconcentrated and decentralized services to the rest of the territory. Indeed, good coordination between transport and local public services can help avoid any possible fragmentation in the social fabric or, in a case of its presence in society, to mitigate the gap. Maintaining this coordination is a big step towards social inclusion and egalitarianism.

One can conclude, in this sense, that public transport constitutes a priority solution in both developing and developed countries and a solution to the urbanistic, economic and social issues. In fact, public transport consumes less space and energy, pollutes less, costs less for the user (cost of the journey) and for the community (infrastructure, social cost).

Thus, the interaction between transport and urbanization and its impact on the social network is indeed a complex subject. However, it is not an insurmountable frontier for action and reflection. The multiplicity of the above-mentioned issues and their complexity show that any action regarding transport should take into consideration both urban planning, economic and social aspects. This is what reflects its multi-dimensionality.

---

[20] Bailly and Heurgon [2] Nouveaux rythmes urbains: quels transports ? Rapport du Conseil national des transports. France: Éditions de l'aube, p.98.

[21] Bailly and Heurgon [2] Op., Cit, P. 96.

11.4.4 The Responsibility of Public Actors and Delegated Companies in the Segregation Process

According to the specifications issued by the territorial authorities[22] for the delegated management of the urban transport service, and more precisely in article 5: the delegator aims to restructure, improve the quality and attractiveness of the Public Transport Service by Bus and increasing its capacity, efficiency and performance. This means that the delegated company has as an obligation to provide good management of urban transport in the perimeter which has been allocated to it, namely the center and its periphery. On the other hand, poor management of urban public transport resources will directly impact its users.

In our case, the population most vulnerable to this mismanagement are those who do not have a choice between different alternatives of public transport means. The center of the conurbation of Rabat-Témara has a multitude and a satisfactory frequency of public transport (Tramway, bus, large Taxi and small taxi). The peripheral municipalities (Ain attiq, Mers el khir, Tamesna, El menzeh), however, have only the bus as the predominant means of transport and large taxis with a reduced quantity. This means that poor management of buses would impact their freedom of movement and can eventually lead to a feeling of segregation.

The diagnostic report "Employment, transport and mobility"[23] Nduwima [15] of 2020 of the National Institute of Planning and Urbanism of Morocco, highlights the difficulties to which the population of the periphery of the region are exposed to with respect to the lack of frequency, the absence of a fixed timetable and the delay experienced by the bus in these areas. These behaviors go against the operating conditions of the contract relating to delegated management (Articles 14, 15 and 16).[24]

However, the delegating authority, namely the prefecture and the wilaya, is also responsible for the fact that they do not exercise regular control and monitoring of the delegated company and do not penalize in the event of non-compliance under their obligation in accordance with Articles 17 and 18 of Law No. 54–05[25] on the delegated management of public services.

---

[22] Portail national des collectivités territoriales https://www.collectivites-territoriales.gov.ma.

[23] Nduwima [15] Emploi, transport et mobilité. Rapport de diagnostic. Maroc: INAU.

[24] https://www.collectivites-territoriales.gov.ma/fr/publications/contrat-type-relatif-la-gestion-des-lignes-du-transport-urbain-par-autobus.

[25] Dahir n° 1-06-15 du 15 moharrem 1427 [8] portant promulgation de la loi n° 54-05 relative à la gestion déléguée des services publics.

11.5 Conclusions

The conclusion that may be projected from this analysis is that despite the urban, economic and social challenges, public transport remains a priority solution in both developing and developed countries. Public transport consumes less space and energy, pollutes less, costs less for the user (cost of the trip) and for the community (infrastructure, social cost). However, the concerned authorities must show a real political and administrative will in order to work for good management that will allow this process to succeed over time.

In short, the interaction between transport and urbanization and its impact on the social fabric is indeed a complex subject. However, it is not an insurmountable frontier for action and reflection. The multiplicity of the aforementioned issues and their complexity show that any action in the field of transport should take into consideration both urban planning, economic and social aspects. This is what reflects its multi-dimensionality.

The main difficulty encountered during the development of this article is to prove the link or the correlation that may exist between the transport system and the phenomenon of peri-urbanization within the Rabat Temara agglomeration and its peripheral areas. Indeed, measuring the phenomenon of peri-urbanization and assessing the performance of the transport system requires taking into consideration a set of criteria, endogenous and exogenous, both quantitative and qualitative to reflect a tangible reality.

However, the availability and access to certain data remains difficult in absolute terms. Nevertheless, this scientific proposal was an opportunity to open up new avenues of reflection and to propose the construction of composite indices relating to peri-urbanization and the transport system to better appreciate the link that can connect them.

References

1. Ariotti, A. (2011) Les effets de la périurbanisation amplifient les inégalités sociales. France : Les Cahiers du Développement Social Urbain À La Voulte-sur-Rhône.
2. Bailly, J. P., Heurgon, E. (2001) Nouveaux rythmes urbains : quels transports ? Rapport du Conseil national des transports. France : Éditions de l'aube, p. 98.
3. Beaujeu-Garnier, J. (1995) Géographie urbaine. France : Armand Colin, p. 25.
4. Bogaert J, Halleux JM (2015) Territoires périurbains : développement, enjeux et perspectives dans les pays du Sud. Presses agronomiques de Gembloux, Belgique
5. Certu. (2008) Mobilités et transports, Fiche n°1, Comment élaborer des stratégies de mobilité durable dans les villes des pays en développement ? France : Note de synthèse, p. 8.
6. CETUR. (1994) Les enjeux des politiques de déplacement dans une stratégie urbaine. France : Cetur, p. 46.
7. Charlot, S.et al. (2006) Périurbanisation, ségrégation spatiale et accès aux services publics. France : Ministère de l'Equipement, des transports, de l'Aménagement du territoire, du Tourisme et de la Mer.

8. Dahir n° 1–06–15 du 15 moharrem 1427 (14 février 2006) portant promulgation de la loi n° 54–05 relative à la gestion déléguée des services publics.
9. Ghezal I, Bouchemal S (2014) Durabilité et périurbanisation : cas de la ville de Hamma Bouzian. Mémoire pour l'obtention du diplôme de Master en urbanisme. Université d'Oum El Bouaghi, Algérie
10. Haut-Commissariat au Plan. (2005) Recensement générale de la population et de l'habitat de 2004, Maroc.
11. Haut-Commissariat au Plan. (2015) Recensement générale de la population et de l'habitat de 2014, Maroc.
12. J Masson, S. (2000) Les interactions entre système de transport et système de localisation en milieu urbain et leur modélisation. France : Thèse de doctorat en sciences économiques, Mention Economie des transports. Lyon 2, p.4.
13. Merlin P (1984) La Planification des transports urbains. Masson et Cie, Paris, p 46
14. Ministère de l'urbanisme et de l'aménagement du territoire. (2016) SDAU 2040 Rabat-Salé-Témara et sa zone périphérique. Maroc : Direction de l'urbanism, p.17.
15. Nduwima, C., Ouhnini, M., Takoune, O. (2020) Emploi, transport et mobilité. Rapport de diagnostic. Rabat : INAU.
16. Portail national des collectivités territoriales https://www.collectivites-territoriales.gov.ma/fr/publications/contrat-type-relatif-la-gestion-des-lignes-du-transport-urbain-par-autobus.
17. Roux, E., Vanier, M. (2008) La périurbanisation : problématiques et perspectives. France : La Documentation française. DIACT, p. 87.
18. Zahiri, M. (2005) Urbanisation, déplacements et transports urbain dans le grand Agadir. Mémoire de 3ème cycle pour l'obtention des études supérieures en aménagement et urbanisme. Agadir : Université Ibn Zohr, p.6.

Chapter 12
Location Choices of Micro Creative Enterprises in China: Evidence from Two Creative Clusters in Shanghai

Zhu Qian

12.1 Introduction

Creative industries have been regarded as one of the most dynamic sectors in the backdrop of economic globalization and booming information and communication technologies [1, 2]. The rise of creative industries has contributed to changes in economic, cultural, social, and urban spatial aspects of our daily life [3, 4]. In the spatial dimension, empirical evidence suggests that creative industries have no propensity to be evenly distributed across space but concentrate in agglomerations to form the so-called creative clusters [5–9]. The formation and development of creative clusters can be either spontaneous or policy driven. In both situations, the growth of creative clusters is the consequence of joint forces of the market, local government, and the creative class, all of which have played critical roles in creative cluster development [10, 11]. Research has suggested that the spatial clustering of enterprises in same sector or industry can help these enterprises to get access to specialized resources and services [1, 12, 13]. Meanwhile, it is argued that traditional economic approaches cannot fully explain the location choice behavior of creative enterprises because institutional environments, industrial characteristics, urban amenities, degree of tolerance, and enterprise sizes all matter in creative enterprises' location choice [14–18].

Micro enterprise is a subset of small- and medium-sized enterprises with nine or less employees [19]. Some of these enterprises may never have intention to grow larger, but others are start-ups with ambitious plans to expand and become influential in the future [20]. Micro enterprises provide employment opportunities in the market and their influences in economic and social aspects have constantly increased [21]. It has been found that micro enterprises make their location choice differently from what medium- and large-sized enterprises do [22].

Z. Qian (✉)
School of Planning, University of Waterloo, Waterloo, ON N2L 3G1, Canada
e-mail: z3qian@uwaterloo.ca

© The Author(s), under exclusive license to Springer Nature Singapore Pte Ltd. 2023
L. Zhang et al. (eds.), *The City in an Era of Cascading Risks*, City Development: Issues and Best Practices, https://doi.org/10.1007/978-981-99-2050-1_12

This research aims to explore the primary factors behind micro creative enterprises' (MCE) location choice, through a study of two micro creative enterprise clusters in central Shanghai, China. It identifies the roles of the market, local authorities, the creative class in the formation and growth of micro creative clusters. The study explores the factors of MCE's location choice from the economic, institutional, and creative class perspectives and seeks feasible improvements for the sustainable growth of micro creative clusters in Chinese cities. Little research has evaluated the factors that impact micro creative enterprises' location choices and their weights at the neighborhood level, especially from the enterprise's perspective.

Shanghai has long been recognized as one of the most cosmopolitan, open, and vibrant cities in China, with a considerably large number of creative enterprises [23]. It is also one of the first Chinese cities that have adaptively reused derelict industrial facilities in their old towns to accommodate micro creative enterprises, as part of the city's strategic response to the economic restructuring from a manufacturing dominated economy to a service oriented one. Shanghai has been home to 87 officially certified creative clusters [24]. However, Shanghai's development of creative industry has sometimes been questioned as the city views creative clusters as a panacea for its service sector growth and for reuse of obsolete industrial facilities [25].

The research fieldwork, consisting of field observation, interviews, and questionnaires, was carried out in April 2019. Local planning context, policies, and implementations in relation to creative clusters, creative industries, and micro enterprises were reviewed to understand how these factors impacted the development of local creative clusters. Field observation of the two study areas was performed to explore and compare their urban landscape, physical environment, spatial structure, as well as creative industry milieu. Interviews with management sectors of creative clusters were conducted to advance the understanding of development process, management mechanism, and cluster sustainability. A questionnaire about how micro creative enterprises chose their locations were distributed to participants in the two study areas–a spontaneous creative cluster M50 and a policy-led creative cluster the Bridge 8. The first part of the questionnaires collected information about enterprise types and sizes, employment, entrepreneur education attainments and their university locations. The second part of the questionnaires gauged location determinants that included 12 factors in economic, institutional, and creative aspects. Economic factors contained agglomeration effects, rent costs, geographical proximity to universities, research institutes, and technological enterprises. Institutional factors involved policy incentives and supports for micro enterprises and creative industries. Creative factors consisted of urban and architectural aesthetics, openness and tolerance, cultural diversity, mixed-use land development, and human-scaled spatial structure and design. Five-Point Likert Scale method was used with the choices of 'most unimportant', 'somewhat unimportant', 'moderate', 'somewhat important', and 'most important'. The third part concerned the main challenges faced by micro creative enterprises and the recommendations. Descriptive analyses were employed to examine the weight and significance of various factors. Factors were compared not only between the development patterns of the two types of creative clusters but also between two sub-sectors of creative industries to reveal the differences. Some of the proposed recommendations were discussed with management staffs during the interviews to assess their feasibility and potential implementation. A total of 51 valid questionnaire responses in M50 and a total of 47 valid responses in the Bridge 8 were recorded for the analysis.

12.2 Micro Creative Class, Stakeholders, and Location Choice

12.2.1 Spontaneous and Policy-Led Creative Cluster Developments

Creative industry has its origin in individual creativity, skill and talent, and demonstrates their potential for wealth and job creation through the generation and exploitation of intellectual property in sectors such as advertising and marketing, architecture, crafts, design, media, IT, computer services, publishing, cultural institutes, and arts [26]. A place that accommodates a wide range of individuals and enterprises that engage with creative industries is creative cluster, where vibrant environment encourages knowledge exchange and stimulates individuals' creativity [7, 27].

A spontaneous and bottom-up creative industry cluster typically uses derelict industrial areas in old town revitalization [3, 28]. Creative class rents disused buildings and renovates them for both work and living at very affordable costs [10]. Spontaneous creative clusters usually engender and grow in a tolerant culture and environment with a supportive market. Gradually, local authorities intervene and integrate spontaneous creative cluster development into their broad urban strategy and enable creative clusters to gain competitive advantages [6, 29]. A policy-led creative cluster development is typically a top-down process in which local authorities intentionally create and sustain creative clusters through various policy initiatives as part of their urban strategies [29, 30]. Policy-led creative clusters may face issues such as over-commercialization, lack of creativity, low degree of openness, and high entry barrier [31, 32]. Both spontaneous and policy-led creative cluster developments may overlook the importance of forming value chains and embedded ties [33] and place-specific features [31].

12.2.2 Local Authority, Creative Class, and the Market

Creative cluster development is often associated with urban revitalization [1, 29, 34]. Demands for creative industry growth entice real estate developers to renovate disused industrial heritage into creative spaces for commercial gains and possible long-term benefits [11]. The clustering of creative enterprises contributes to generating competitive advantages through enhancing the productivity of enterprises in the cluster and stimulating their innovation and entrepreneurship [35, 36].

Local authorities have two major motivations for creative cluster development–urban revitalization and new economy growth [3, 33]. Local governments approach urban revitalization for city image building, cultural diversification, openness and sustainability [11]. Local authority's engagement with creative cluster development typically entails urban development strategies that incorporate creative cluster components, legislations and regulations for creative clusters and related industries, comprehensive planning, public engagement, direct supports for creative cluster development, and institutional and administrative interventions [37, 38]. In addition, local authorities often play the role of coordinating issues among other agents in the market [6].

Creative class, especially in spontaneous creative cluster formation, tends to actively engage with architectural heritage preservation of disused built environment, aiming to foster an open, dynamic, personal, and professional creative milieu that "contains the necessary preconditions in terms of 'hard' and 'soft' infrastructure to generate a flow of ideas and inventions ... and as a consequence contributes to economic success" [39].

12.2.3 Location Choice in Theoretical and Empirical Discourse

Hayter [40] categorizes three strands of location choice theories. In the neoclassical theory, an enterprise's location decision is influenced by factors that can create expected profits such as agglomeration economies, market size, labour costs, and transportation. The institutional theory focuses on the role of institutional instruments when examining an enterprise's location decision. The behavioral theory highlights the significance of internal and entrepreneurial factors and maintains that small enterprises sometimes have to make their location decisions in an uncertain environment with limited locational information and time, so their location choices are often built on subjective factors such as geographical origin, graduation place, and other personal experiences.

Micro creative enterprises, like any other enterprises, choose locations to help maximize their profits [41]. Economic globalization makes location choice consider a diverse list of location determinants such as economies of scale, natural and urban amenities, rent cost, logistics, human capital, taxes, open and tolerant environment [5, 41]. Densely populated urban regions are likely to attract creative people largely because of their cosmopolitan amenities, knowledge intensive sectors, specialized labor sources, human capital, and tolerant social and institutional environment [9, 42].

Besides benefits from industrial network, branding effect, suppliers and customers, clustering of micro creative enterprises facilitates the spillover of tacit knowledge and stimulates innovation [1, 43]. Transaction costs can be reduced when micro creative enterprises cluster in places for better coordination, collaboration, and mutual trust. As a knowledge-intensive sector, micro creative enterprises prefer productive amenities with ample skilled labor resources, specialized enterprises, universities, institutes, and quality infrastructure [42].

Micro creative enterprises' location choices are impacted by subjective elements that may not necessarily influence larger enterprises. For instance, owners of micro creative enterprises are more likely to begin their businesses in their geographical origins or places where they graduate to retain and take advantage of existing social networks [44]. University resources can facilitate commercialization of knowledge and help generate new business ideas [45]. Cheap rent is another economic factor in micro creative enterprise location choice, especially for start-ups. Government policy incentives such as loans, grants, subsidies, tax refunds and abatements are other major factors that micro creative enterprises typically consider for their location choice [46–48].

Historic built environment and iconic infrastructure can be contributive to the cultivation of place branding [49]. High accessibility, good walkability, and efficient transit services are often assets. However, favored built environment might trigger property value increase and gentrification, which eventually lead to disamenity for micro creative enterprises with limited financial resources [50]. Open, vibrant, multi-cultural and inclusive places that demonstrate tolerance and openness to immigrants, foreigners, and sexual and racial minorities are positive social factors to most micro creative enterprises [42].

12.3 Two Micro Creative Clusters in Shanghai

12.3.1 Creative Industries in Shanghai

Creative industry development has been closely connected to Shanghai's old city revitalization and economic restructuring since the 1990s [38]. In Shanghai's old districts, a growing number of disused industrial structures were leased for other types of uses at low rental prices. A few creative clusters were gradually formed in a spontaneous way and have become well known, such as M50, Tianzifang, and Sihang Warehouse. These art-oriented creative industry communities adopted participatory planning in the early stage of development by involving local stakeholders such as creative class, residents, property owners, investors, and NGOs [51]. Since the 2000s, the municipal government has analyzed the needs of creative industries and their preferences, as part of their assessments of the regional development conditions for urban strategy, and subsequently, defined specific geographical areas for creative industries. The local and district governments have formulated development strategies, regulations, and incentives for these designated creative clusters. In 2004, Shanghai established Shanghai Creative Industry Center, a semi-governmental organization, to assist the municipal government to formulate development plans and strategies for creative industries. In early 2005, Shanghai Commission of Economy and Informatization officially certified the first 18 creative clusters, and there are now 87 certified creative clusters in Shanghai. Shanghai joined the UNESCO Creative Cities Network in 2010 and demonstrated the city's ambition to place creative industries at the core of the city's development. Shanghai Cultural and Creative Industry Promotion Office, also established in 2010, exercises strong institutional power of resource integration, inter-sectoral coordination, and international cooperation in creative industries [52]. The *Shanghai Master Plan (2017−2035)* articulates that the city will continue to provide micro enterprises with financing, consulting, training, and other professional services, as well as relevant infrastructures in order to create a more competitive environment for entrepreneurship [53].

Creative industries in Shanghai are formally categorized into Technology-based Industries (TBIs), Service-based Advisory Industries (SBAIs), and Cultural, Art and Fashion Industries (CAFIs). The majority of Shanghai's creative clusters are located in central areas, enclosed by the Inner Elevated Ring Road and the Huangpu River, where the cosmopolitan built environment and the openness are favored by the creative class. The certified creative clusters all have a public–private partnered management sector to oversee the daily operation of creative enterprises on behalf of local authorities, who play the roles of mediator and facilitator among stakeholders in policy implementation and government grant allocation [11].

12.3.2 Two Creative Industry Clusters-M50 and the Bridge 8

Located within the old city of Shanghai, both M50 and the Bridge 8 were certified in the first around creative cluster accreditation by the local government. These two clusters have enjoyed good place-based reputation and reflected the typical spatial characteristics of Shanghai's creative clusters. Used to be the site of a state-owned factory known as Shanghai Chunming Slub Mill, M50 (or 50 Moganshan Road) is located in an old industrial area along the Suzhou Creek in Putuo District. In the late 1990s, the factory was closed down during the restructuring of Shanghai's textile industry. In 2001, the ownership of these old industrial buildings was transferred to ShangTex, a large state-owned textile and garment group. Then Shanghai Chunming Slub Mill was renamed to M50 which later became a creative industry cluster. The initial growth of M50 was unplanned and spontaneous. When local authorities recognized the potential of developing creative cluster, they started to intervene by institutional powers in 2005 [11]. M50 has become a contemporary art community, accommodating over 130 domestic and overseas artists and creative enterprises in its 25 buildings constructed in a timespan from the 1930s to 1990s [25].

The Bridge 8 (or Ba Hao Qiao in Chinese), located in the former French Concession of Huangpu District, used to be the site of Shanghai Automobile Brake Factory in Middle Jianguo Road. The ownership of the factory was transferred to Shanghai Huaqing Investment and Life-style Center Holdings and the place was soon reconstructed and renamed to the Bridge 8 in the end of 2004. Initiated by the joint efforts of Shanghai Municipal and Huangpu District governments, the development of the Bridge 8 has envisaged creative industries as advanced business services in upscale office environment in reconstructing and renovating disused structures. Today, the Bridge 8 consists of two areas with four sub-clusters named the Bridge 8 Phases I, II, III, and IV. Among the four phases, Phase I is the earliest and best known. The Bridge 8 has attracted a diversity of creative enterprises in architectural design, media, advertising, fashion design, etc. Interestingly, the name of the Bridge 8 was originally from the connections between buildings, but now has become a concept and branding sign of the place (Fig. 12.1).

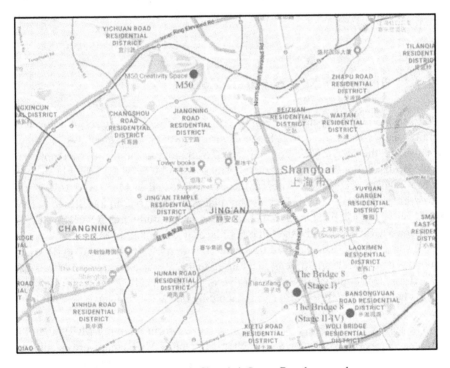

Fig. 12.1 Locations of two study areas in Shanghai. *Source* Based on google map

12.4 Research Findings

12.4.1 Micro Creative Enterprise Types

The survey shows that M50 accommodated all eight types of creative industries but crafts had the most significant share, with nearly 40% of all MCE survey participants. The share of CAFIs was a heavily weighted 84% in M50, while only 16% of MCEs were SBAIs. In contrast, there was no single dominating creative industry in the Bridge 8. Approximately 57% of all MCEs were SBAIs and 43% of all MCEs were CAFIs. M50 was a CAFI dominated cluster because M50 became a creative cluster after avant-garde artists moved into the old industrial site and made great efforts to renovate the place. And the tradition remained over the years. The mature business mode, along with well-developed infrastructure, attracted various types of MCEs in the Bridge 8.

12.4.2 Enterprise Size, Origin, and Age

MCEs in M50 typically had 2–4 employees, many of which were self-employed or family businesses. In the Bridge 8, the majority of MCEs had 6 employees or more. On average, the size of MCEs in M50 was 5.6 employees and it was 7.8 employees in the Bridge 8. For the origins of their businesses, 38 MCEs started their businesses in M50, which accounted for approximately 75% of all in M50. And 16% of the surveyed MCEs were established in other places in Shanghai. In the Bridge 8, 51% of the surveyed MCEs started their businesses in the study area. Over 44% of the surveyed MCEs were established in other places in Shanghai. In M50, roughly 31% of the surveyed MCEs had stayed for one to four years, 20% of the surveyed MCEs had stayed for five to eight years, and more than 25% of the surveyed MCEs had stayed for over eight years. In the Bridge 8, approximately 70% of the surveyed MCEs had stayed for one to four years and only one MCE had stayed for over five years.

The survey indicates that M50 had a better record in retaining MCEs than the Bridge 8. However, the Bridge 8 was better in attracting MCEs that were established in other places. When comparing with MCEs in M50, one may argue that those in the Bridge 8 had higher possibility of changing their workplaces. The findings suggest that in both study areas some MCEs suffered from substandard management. And overdevelopment, homogenization, and other external reasons could result in the high location mobility.

12.4.3 Entrepreneur Educational Attainment and Location

The educational attainments of entrepreneurs were evenly distributed across three levels (college, bachelor's degree, and master's degree and above) in M50. Entrepreneurs who obtained a bachelor's degree in the Bridge 8 had a similar proportion as in M50, but the Bridge 8 had up to 60% of its entrepreneurs with a master's degree or above. Only 6% of the surveyed entrepreneurs in M50 graduated in Shanghai. In contrast, entrepreneurs who graduated outside Shanghai had the proportion of over 80%. Seven entrepreneurs (14%) had studied abroad before they started their businesses in M50. In the Bridge 8, 28% of the surveyed entrepreneurs gradated in Shanghai and 32% graduated in other places. It is important to note that entrepreneurs who obtained their degrees overseas accounted for over 40% in the Bridge 8. Mellander and Florida [4] maintain that it is more crucial to gauge what people do (their occupations) than what they study (their educational attainment). The interviews reveal that entrepreneurs with higher educational attainments and overseas educational experience were more likely to possess a global outlook which contributed to a more international, tolerant, and professional creative milieu.

Entrepreneurs who were born in Shanghai shared a small proportion of 12% in both study areas. This result aligns with the field observation which showed that most creative people communicated in Mandarin instead of Shanghainese, a local dialect. It is argued that the more non-locals the areas had, the more tolerant the areas were. Apart from that, one third of the surveyed entrepreneurs in M50 were born outside Shanghai. They moved to Shanghai and then changed their *Hukou* (household registration) status. In the Bridge 8, this cohort of entrepreneurs were about half of all survey participants (Fig. 12.2). The results reflect the attractiveness of Shanghai to outsiders, especially entrepreneurs and young graduates in creative industries [10]. Moreover, social ties affected entrepreneurs' location decisions. Many chose M50 because of the suggestions from their friends and colleagues.

12.4.4 Main Location Choice Determinants

In determining on the influences of various factors, a Five-point Likert Scale method was used and scores ranging from 1 to 5 were given for 'most unimportant', 'somewhat unimportant', 'moderate', 'somewhat important', and 'most important', respectively. In M50, *Urban and architectural aesthetics* was the most critical location determinant, which had the highest mean score (4.4 out of 5), followed by *Branding and place-based reputation* (4.3) and *Agglomeration effect of an industry* (4.2). The location choice of MCEs in M50 was mainly influenced by economic and creative factors rather than institutional factors (Fig. 12.3). The mean scores of the influential factors were relatively close to each other in the Bridge 8, ranging between 4.1 and 3.3. It suggests that the location choice behavior of most MCEs in the Bridge 8 was equally influenced by economic, institutional, and creative factors

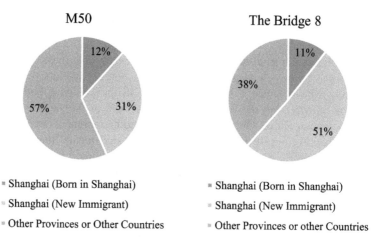

Fig. 12.2 Entrepreneur's *Hukou* status by percentage—M50 (left) and the Bridge 8 (right)

(Fig. 12.4). Serra [9] stresses that the spatial agglomeration of economic activities cannot impact all industries in the same way. The research investigated the differentiation in their location choices of two sub-sectors of creative industries—SBAIs and CAFIs (Fig. 12.5).

About 86% of the surveyed MCEs in the Bridge 8 highlighted that industrial agglomeration effect was one of their main location considerations. Up to 86% of the surveyed MCEs in M50 took branding and place-specific reputation into consideration when they evaluated a potential location. In comparison, the share of this same consideration was smaller (62%) in the Bridge 8 (Figs. 12.6 and 12.7). The authenticity and reputation of M50 has become place-specific reputation that is commonly

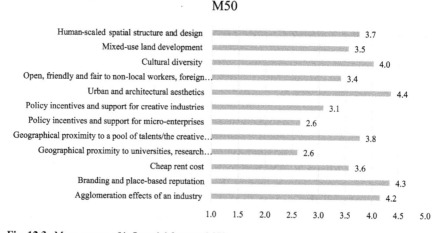

Fig. 12.3 Mean scores of influential factors, M50

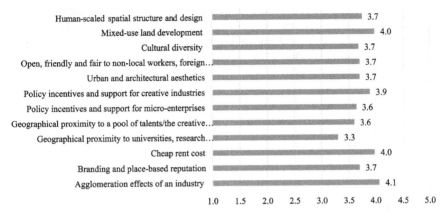

Fig. 12.4 Mean scores of influential factors, the Bridge 8

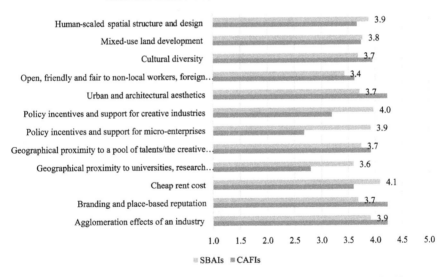

Fig. 12.5 Comparison of the mean scores of influential factors between SBAIs and CAFIs

acknowledged as a catalyst for local economic vitality. Place-specific reputation also helps promote tourism and attract investments. Cheap rent was less attractive for the surveyed MCEs in M50 than those in the Bridge 8. A managerial staff in the Bridge 8 mentioned in an interview that policy-led creative clusters typically received more public subsidies than spontaneous ones, which significantly reduced their financial burdens.

Merely 30% of the surveyed MCEs in M50 thought that *Policy incentives and support for micro enterprises* were critical in their location choice. The MCEs' choices were evenly distributed among four levels, ranging from 'most unimportant' to 'somewhat important'. Thus, *Policy incentives and support for micro enterprises*

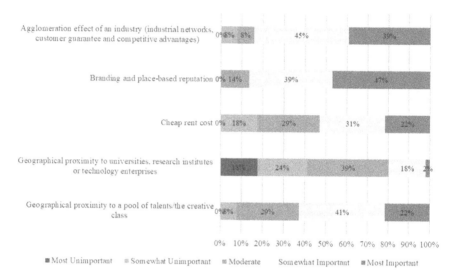

Fig. 12.6 Location choice for economic factors, M50

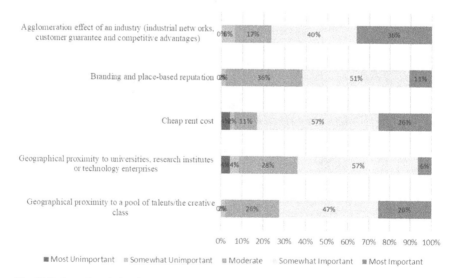

Fig. 12.7 Location choice for economic factors, the Bridge 8

was ranked as one of the least influential factors in M50. Putuo District introduced financial incentives to encourage the development of micro enterprise such as *Putuo District Small and Micro Enterprises Special Credit Loans* and *Putuo District Special Fund for Cultural Industries Development* [54]. Yet, there was no significant difference in choosing between M50 or other places in Putuo District in terms of institutional support. In the Bridge 8, more than half of the surveyed MCEs were reliant on policy incentives and support for both micro enterprises and creative industries.

About 84% of the surveyed MCEs in M50 could be categorized as CAFIs, while 57% of the surveyed MCEs in the Bridge 8 could be categorized as SBAIs. These were the two dominating sub-sectors in each creative cluster and had an impact on MCEs' location choices. SBAIs are highly reliant on context-specific services that are based on professional knowledge and ICT. M50 is a CAFIs-dominated cluster, for which O'Connor and Gu's [10] research maintains that entrepreneurs in CAFIs (such as visual art, fashion, music and design) are more 'footloose', and thereby are reluctant to have strong connections with local authorities. In addition, creative people in CAFIs usually are involved in 'individualized' creativity, while SBAIs are often project-oriented, featuring 'collective' creativity [9, 16].

An open, tolerant, and diverse environment matters. About 74% of the surveyed MCEs in M50 and 60% of those in the Bridge 8 valued cultural diversity of a place, including vibrant cultural events and quality facilities. M50 held a wide range of activities, such as university student entrepreneurship fairs, children art and creativity competitions, modern dramas, etc. The survey also reveals that the MCEs in the Bridge 8 had higher expectation for a wide range of urban amenities than those in M50. More than 80% of the surveyed MCEs in the Bridge 8 highly valued a mixed-use and amenity-rich community. In terms of the built environment, about 75% of the surveyed MCEs in M50 and 62% of the surveyed MCEs in the Bridge 8 took walkability, accessibility, and open space into account when they considered enterprise locations (Figs.12.8 and 12.9).

MCEs in both study areas were satisfied with architectural design, urban amenities, and policy incentives. Comparing with those in the Bridge 8, the surveyed MCEs in M50 hoped to gain more public engagement in the development process and more accessible public and green spaces for social interaction and relaxation. Nearly half of the surveyed MCEs lamented that since 2005, when M50 was officially certified by the local authorities, the engagement of creative people had gradually diminished. The original bottom-up development was replaced by the increasingly tight institutional controls [25]. Consequently, micro creative enterprises in M50 seemed to become tenants. The situation was different in the Bridge 8, a top-down planned creative cluster, where enterprises had no direct linkage with the place in the initial development stage. Creative people had less place-associated identity than those in M50, and simply acted as business tenants. Drinkwater and Platt [3] highlight that institutional factors may have a positive impact on the clustering of creative industries in some places but may not in others. Creative cluster development ought to place people, instead of the built environment, at the core of planning and development processes. During the fieldwork, many MCEs complained about rising rental costs—45% of the survey participants in M50 and 55% in the Bridge 8. Unaffordable

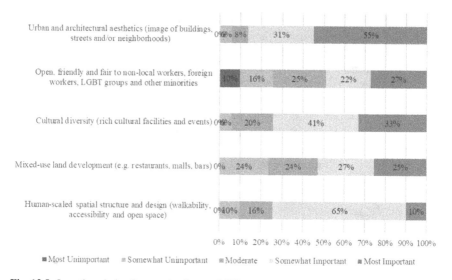

Fig. 12.8 Location choice for creative factors, M50

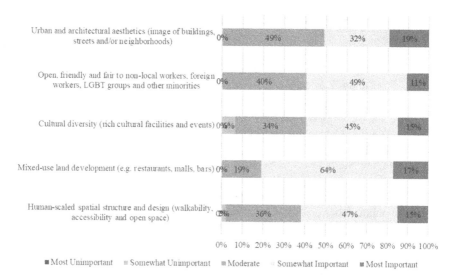

Fig. 12.9 Location choice for creative factors, the Bridge 8

rental costs are one of the major challenges that micro enterprises are faced, which can discourage innovation and creativity [11].

In the Bridge 8, the share of non-creative enterprises accounted for approximately 30%, verified by both the field observation and the interviews. There were a number of restaurants, cafés, and retails that had nothing to do with creative industry. There has been critiques about commercial development disguised by the name of creative cluster, illegal reconstruction, high vacancy rate, non-creative enterprises, and other maladministration problems [38]. Zou and Liu [55] point out that many creative cluster management sectors merely maintain an owner-tenant relationship with their creative enterprises and make limited efforts to seek long-term sustainable development of creative clusters. Creative clusters have struggled with high vacancy rates in Shanghai. As observed in the field, buildings in the Bridge 8 Phases II and IV had high vacancy, caused by the lack of creativity, overbuilding, and competition against other local creative clusters. In response, the Bridge 8 had built a business partnership with a Hong Kong based enterprise. Their joint office project *the Desk Bridge 8* in Phase IV aimed to form an efficient knowledge sharing platform and build an inclusive community for medium, small and micro creative enterprises.

12.5 Conclusions

This research aims to understand the roles of the market, local authorities, and the creative class in the development of Shanghai's creative clusters, and to explore the factors impacting MCEs' location choice in economic, institutional, and creative aspects. Local authority and the creative class sometimes had different motivations and pursued their specific goals, each having ongoing influence on the sustainability of creative cluster development. Meanwhile, their influences varied, depending on the types of development patterns. In the spontaneous development pattern, the creative class played a central role in the initial development stage. The agglomeration of creative people fostered an open, dynamic, personal, and professional creative milieu. Creative people demonstrated strong interests in historical preservation of their places. In the policy-led development pattern, creative clusters had been well planned and designed before creative people moved in. In the spontaneous development pattern, local authorities showed tolerance to the bottom-up planning by creative people at the initial development stage due to inadequate ability of local authorities to impose strong control on urban revitalization. When local policymakers and administrators realized the importance of creative clusters, they then started to intervene in creative cluster development by acting as planners, sponsors, and even investors.

Both development patterns effectively reshaped urban space. In M50, the creative class spontaneously renovated warehouses and factory facilities by applying their aesthetic knowledge and preferences. By contrast, buildings in the Bridge 8 were redesigned by professionals commissioned by the local government. There, the style of renovated buildings was contemporary, upscale, and with a global vision. Economic motivations influenced MCEs' location choices. Most of MCEs tended to agglomerate in specific places to benefit from a variety of industrial agglomeration effects, including industrial networks, customer sources, and competitive advantages. Such location choice behavior accorded with the general theory of industrial location choice. However, the spatial agglomeration of economic activities does not impact all industries in the same way. The research finds that creative factors and institutional factors also had impacts on MCEs' location behavior due to their industrial characteristics and enterprise sizes. Furthermore, though most of the proposed location determinants were shared by both micro creative enterprises and general creative enterprises, micro creative enterprises preferred to be in well-known places and/or places where they had strong social ties. These places can help MCEs raise enterprise recognition, reduce enterprise costs, and resist market risks.

Location determinants of MCEs in M50 were not exactly the same as those in the Bridge 8. For MCEs in M50, apart from industrial agglomeration effects and branding and place-based reputation, urban and architectural aesthetics were viewed as another significant influential factor, which could be explained by the dominant industries in M50–CAFIs. By contrast, MCEs in the Bridge 8 tended to choose their locations based on comprehensive consideration in economic, institutional, and creative aspects. MCEs generally had a positive attitude towards the study areas in terms of openness and tolerance, institutional support, and urban spatial design. Both development patterns were able to attract, retain, and nurture MCEs in some ways. However, many MCEs were confronted with development challenges, mainly in economic and creative aspects. The sustainability of a creative cluster is reliant on not merely spatial coexistence of different creative enterprises, but the ability of generating and sustaining economic prosperity that will continuously attract others into the cluster [3]. Both study areas should pay attention to at least three main aspects—improving public engagement, cultivating creative milieu, and strengthening cooperation and linkage, all of which require a long-term process. Policy incentives should be more specific and flexible for various sub-sectors and sizes in creative industries. The interviews with the management staff in the two study clusters suggest that both creative clusters have been seeking feasible and sustainable solutions to create and sustain long-term progress and place people rather than the built environment at the center of creative cluster planning process. The clustering of creative industries needs to shift from the conventional focus on infrastructure and environment improvements to the attention to technology and knowledge spillover strategies. These changes will support creative clusters in Chinese cities to continually reinvent themselves in changing socioeconomic conditions.

References

1. Brinkhoff S (2006) Spatial concentration of creative industries in Los Angeles. NEURUS-Network of European and US Regional and Urban Studies.
2. United Nations Conference on Trade and Development (2019) UNCTAD's work on the creative economy. Retrieved from https://unctad.org/en/Pages/DITC/CreativeEconomy/Creative-Economy.aspx.
3. Drinkwater BD, Platt S (2016) Urban development process and creative clustering: the film industry in Soho and Beyoğlu. Urban Design International 21(2):151–174
4. Mellander C, Florida R (2007) The creative class or human capital? explaining regional development in Sweden (No. 79). Royal Institute of Technology, CESIS-Centre of Excellence for Science and Innovation Studies.
5. Florida R (2002) The economic geography of talent. Ann Assoc Am Geogr 92(4):743–755

6. Gong H, Hassink R (2017) Exploring the clustering of creative industries. Eur Plan Stud 25(4):583–600
7. Ma R, Shen Y (2010) The research progress and problem of creative industrial district's theoretical study in China. World Reg Stud 2. (In Chinese).
8. Scott AJ (2006) Creative cities: conceptual issues and policy questions. J Urban Aff 28(1):1–17
9. Serra DS (2016) Location determinants of creative industries' firms in Spain. Investig Reg-J Reg Res 34(34): 23–48.
10. O'Connor J, Gu X (2014) Creative industry clusters in Shanghai: a success story? Int J Cult Policy 20(1):1–20
11. Zielke P, Waibel M (2014) Comparative urban governance of developing creative spaces in China. Habitat Int 41:99–107
12. Hanson GH (2000) Firms, workers, and the geographic concentration of economic activity. In: The Oxford Handbook of Economic Geography. pp 477–494.
13. Zhu H (2008) The Impact of the overseas creative industries on China. Commer Res 10(5):56–59 (In Chinese)
14. Chu J (2009) A study on spatial difference of the creative industrial zones in Shanghai. Econ Geogr 29(1). https://doi.org/10.1145/3132847.3132886. (In Chinese).
15. Clifton N (2008) The "creative class" in the UK: an initial analysis. Geogr Ann, Ser B: Hum Geogr 90(1):63–82
16. Drake G (2003) "This place gives me space": place and creativity in the creative industries. Geoforum 34(4):511–524
17. Rao Y, Dai D (2017) Creative class concentrations in Shanghai, China: what is the role of neighborhood social tolerance and life quality supportive conditions? Soc Indic Res 132(3):1237–1246
18. Tschang FT, Vang J (2008) Explaining the spatial organization of creative industries: the case of the U.S. videogames industry. Entrepreneurship and Innovation-Organizations, Institutions, Systems and Regions, pp 1–38.
19. OECD (2018) Enterprises by business size. Retrieved from https://data.oecd.org/entrepreneur/enterprises-by-business-size.htm.
20. Papadaki E, Chami B (2002) Growth determinants of micro-businesses in Canada. Small Business Policy Branch, Industry Canada.
21. Chan SH, Lin JJ (2013) Financing of micro and small enterprises in China: an exploratory study. Strateg Chang 22:431–446
22. Carod JMA, Antolín MCM (2004) Firm size and geographical aggregation: an empirical appraisal in industrial location. Small Bus Econ 22(3–4):299–312
23. Florida R, Mellander C, Qian H (2008) Creative China the university, tolerance and talent in Chinese regional development. 3(145):1–36
24. Shanghai Commission of Economy and Informatization (2017) The 13th five-year plan for the development of creative industry in Shanghai. Retrieved from http://www.shccio.com/zclm/1695.jhtml. (In Chinese).
25. Gu X (2014) Cultural industries and creative clusters in Shanghai. City Cult Soc 5(3):123–130
26. Department for Culture Media and Sport (2016) Creative industries economic estimates. https://doi.org/10.1177/0002716204270505
27. Florida R (2002) The Rise of the Creative Class. Basic Books, New York
28. Evans G (2009) From cultural quarters to creative clusters–creative spaces in the new city economy. Institute of Urban History, pp 32–59.
29. Zhang X, Hui M (2007) Creative clusters: basic conceptions and international practices. J Jishou Univ 28(4). (In Chinese).
30. Chapain C, De Propris L (2009) Drivers and processes of creative industries in cities and regions. Creat Ind J 2(1).
31. Cohen DE (2014) Seoul's digital media city: a history and 2012 status report on a South Korean digital arts and entertainment ICT cluster. Int J Cult Stud 17(6):557–572
32. Keane M (2009) Great adaptations: China's creative clusters and the new social contract. Continuum 23(2):221–230

33. Flew T (2010) Toward a cultural economic geography of creative industries and urban development: introduction to the special issue on creative industries and urban development. Inf Soc 26(2):85–91
34. He J, Gebhardt H (2014) Space of creative industries: a case study of spatial characteristics of creative clusters in Shanghai. Eur Plan Stud 22(11):2351–2368
35. Marshall A (1890/2009) Principles of economics: unabridged eighth edition. Cosimo, Inc, New York.
36. Porter ME (1998) Clusters and the new economics of competition. Harv Bus Rev 76(6):77–90
37. Lin N (2011) Enlightenment from the policies of Japanese creative industry. J CentL South Univ (Soc Sci) 1:16 (In Chinese)
38. Luan F, Wang H, An Y (2013) The development of creative industrial parks in Shanghai and their spatial features. Urban Plan Forum 2:70–78 (In Chinese)
39. Landry C (2012) The Creative City: a Toolkit for Urban Innovators, 2nd edn. Routledge, London
40. Hayter R (1997) The dynamics of industrial location: the factory, the firm and the production system. Wiley, New York
41. Blakely EJ, Leigh NG (2013) Planning local economic development. Sage
42. Cruz S, Teixeira A (2014) The determinants of spatial location of creative industries start-ups: evidence from Portugal using a discrete choice model approach. In: FEP Working Papers, 546. pp 1–45.
43. Devereux MP, Griffith R, Simpson H (2007) Firm location decisions, regional grants and agglomeration externalities. J Public Econ 91(3–4):413–435
44. Polonyová E, Ondoš S, Ely P (2015) The location choice of graduate entrepreneurs in the United Kingdom. Miscellanea Geographica 19(4):34–43
45. Heblich S, Slavtchev V (2014) Parent universities and the location of academic start-ups. Small Bus Econ 42(1):1–15
46. Berg SH (2015) Creative cluster evolution: the case of the film and TV industries in Seoul, South Korea. Eur Plan Stud 23(10):1993–2008
47. Chandler V (2012) The economic impact of the Canada small business financing program. Small Bus Econ 39(1):253–264
48. Haisch T, Klöpper C (2015) Location choices of the creative class: does tolerance make a difference? J Urban Aff 37(3):233–254
49. Currid E, Williams S (2010) The geography of buzz: art, culture and the social milieu in Los Angeles and New York. J Econ Geogr 10(3):423–451
50. Zandiatashbar A, Hamidi S (2018) Impacts of transit and walking amenities on robust local knowledge economy. Cities 81:161–171
51. He J (2014) Creative industry districts: an analysis of dynamics, networks and implications on creative clusters in Shanghai. Springer, New York
52. United Nations Educational, Scientific and Cultural Organization (2017) Creative cities network-Shanghai. Retrieved from https://en.unesco.org/creative-cities/shanghai.
53. Shanghai Municipal People's Government (2018) Shanghai master plan (2017–2035). Retrieved from http://www.shanghai.gov.cn/newshanghai/xxgkfj/2035002.pdf. (In Chinese).
54. Putuo District Bureau of Finance (2016) Putuo district small and micro enterprises special credit loans in 2016. Retrieved from http://www.shpt.gov.cn/shpt/gkczj-zcyiju/20160217/93454.html. (In Chinese).
55. Zou Y, Liu L (2006) The secret of The Bridge 8's success. China Acad J Electron Publ House. http://ci.nii.ac.jp/naid/40006620894/. (In Chinese).

Chapter 13
On the Spatial Formation Mechanism and Inclusive Development of Tibetan Commodity Streets in Chengdu City

Xiao Qiong

13.1 Introduction

Chengdu city has been an important economic and cultural exchange center in Southwestern China since ancient times as well as an important urban space for ethnic exchanges and integration. In the past 10 years, with the development of urban social economy and the further close social communication among ethnic groups, the number of Tibetan populations in Chengdu has been increasing while Tibetan commodity exchange activities have become more and more frequent. Now a relatively mature Tibetan commodity street with considerable scale and characteristics has been formed around Wuhouci St. in Chengdu. Meanwhile, it is also a concentrated residential area for Tibetan urban mobile population. On the one hand, the formation of Tibetan commodity streets gives new characteristics to the living space of ethnic population today. On the other hand, it also makes Chengdu, a famous historical and cultural city, more inclusive of urban society, cultural diversity, and sustainability of economic development.[1]

---

[1] Supported by 2016 National Social Science Fund "Research on Sustainable Development of Southwest Minzu Characteristic Villages from the Perspective of Economic Anthropology" (No:16BMZ054); 2022 Key Innovation Project of Southwest Minzu University "Survey Study On Urban Integration of the Ethnic Mobile Population in the New Era---A Case Study in Chengdu City, Sichuan Province" (No.13); China Association of Higher Education "The Internal Logic, Realistic dilemma and Path Construction of the Study on Poverty Alleviation through Self-taught Higher Education Examination in Sichuan Ethnic Areas" (No. 21ZKB01).

X. Qiong (✉)
Southwest Minzu University of China, Chengdu, Sichuan, China
e-mail: xiaoqiong71@hotmail.com

13.2 Status of Tibetan Commodity Streets in Chengdu

The Tibetan commodity streets in Chengdu is mainly located in the south of Wuhouci St. and the north of Xiaojiahe in Chengdu. It is a block shaped like a cross mainly composed of Ximianqiaoheng St., Wuhouci East St. and Wuhouciheng St. It is an area where the history and culture of the Three Kingdoms, Sichuan local folk culture and Tibetan traditional culture coexist, with typical characteristics of diversified economy and culture (Fig. 13.1).

13.2.1 Tibetan Commodity Stores in Chengdu

The number of Tibetan commodity stores in the streets are diversified. According to the field survey data in mid-May 2019, there are 307 stores selling Tibetan commodities in Wuhouciheng St., Wuhouci East St. and Ximianqiaoheng St. in Chengdu (See Table 13.1). The commodities sold are roughly divided into Tibetan clothing, Tibetan blankets, Thangka, handmade jewelry, Tibetan books and audio-visual products, Tibetan food, Buddha statues, Buddhist supplies and so on. As for the Tibetan clothes here, there are Tibetan traditional clothes, Tibetan robes, Tibetan clothes with modern design elements which are very popular in Tibetan youth, as well as various

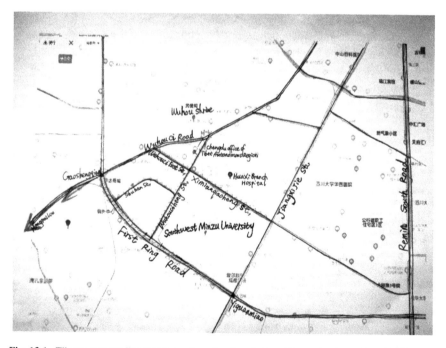

Fig. 13.1 Tibetan commodity streets in chengdu city of china (Drawn by the author, 2021)

Table 13.1 Tibetan commodity Stores in Wuhouciheng St., Wuhouci East St. and Ximianqiaoheng St

No	Types of commodities	Number of stores
1	Buddha statues	88
2	Religion appliances	84
3	Thangka	28
4	Tibetan restaurant	20
5	Lama's robes	19
6	Tibetan clothes	19
7	Tibetan blankets	10
8	Handicrafts	9
9	Ornaments	9
10	Printed materials	5
11	Hotels	5
12	Tibetan incense	4
13	Electric appliance	3
14	Tibetan bookstores	2
15	Stone carving	2
	Total	307

Source The data collected in the field survey by author in May, 2019

robes worn by lamas in Tibetan temples. Daily household appliances include cake-baking machine, butter mixer, milk separator and heater. Buddhist supplies include Hada, colorful prayer flags, handmade Chacha (small Buddha statues made of mud), Buddha beads, bronze mirrors etc. Buddha statues have a wide range of types, sizes and scales with exquisite workmanship and innovative technology.

It is not hard to find from Table 13.1 that there are various types of Tibetan commodities sold in the streets, which related to the Tibetan production, life, and religious beliefs.[2] Obviously, there are the largest number of stores selling Buddhist statues, Tibetan religious utensils, Thangka and Tibetan restaurants. And the number of the stores selling Buddhist statues and religious utensils on the street accounts for as much as 56.3%. It not only reflects that the daily necessities sold in the streets are often indistinguishable from the Buddhist supplies in temples, but also reflects the close relationship between the daily life of the Tibetan people and their Buddhist culture.

---

[2] Due to the rich and diversified supply, most of the commodities in this district are most popular among the Tibetans. Then these are sold to Tibetan areas in Sichuan, Tibet, Qinghai, and all over the country to meet the daily life and religious needs of local Tibetans and Buddhism temples. So, this district is also known as the Tibetan commodity wholesale center.

Table 13.2 Stalls at the night market in wuhouciheng st., wuhouci east st. and ximianqiaoheng st

No	Types of commodities	Number of stalls
1	Handicraft	71
2	Electronic products	19
3	Daily clothing	15
4	Shoes	13
5	Tibetan foods	11
6	Household supplies	6
7	Bags	6
8	Tibetan incense	3
9	Cosmetics	2
10	Popular books	2
11	Toys	1
Total		149

Source The data collected in the field survey by author in May, 2019

13.2.2 Temporary Stalls at the Night Market

Besides the fixed stores, there are many temporary stalls at the night market. Around 7:00 pm in the evening, it is the busy time for the Tibetan people who temporarily live in the family hotels or the neighborhood communities to set up stalls selling small commodities, such as ordinary and cheap jewelry, Buddha beads, clothing, yogurt, Zanba, butter, Yak meat and others[3] (See Table 13.2).

The stalls are not in large scale and the most commodities are from the local Chengdu wholesale center while the seller only earns a certain middle price difference. There are also some traditional handicrafts made by Tibetan themselves and some Tibetan traditional foods brought from their hometowns. Of course, the sellers at night market are almost young and middle-aged Tibetans, and the women make up the majority.

The diversified commodities sold at night market are daily necessities in small scale while the commodity scope and mode of night market is relatively flexible. Then the managing risk faced by the sellers is much smaller. During the epidemic COVID-19 in 2020, The night market in the streets disappeared and consequently, most of the Tibetan sellers at the night market went back their hometown of their own will. The stalls didn't set up again until this early spring.

---

[3] The commodities at the night market mainly meets the daily needs of the Tibetan mobile population in Chengdu, sold rather cheaply and temporarily.

Fig. 13.2 Wuhouciheng St., photo by Xiao Qiong, Tibetan commodity stores, photo by Xiao Qiong

13.3 Characteristics of Tibetan Commodity Streets in Chengdu

13.3.1 Distinctive Ethnic Traditional Culture

There are over 85% stores in the streets selling Tibetan commodities which covers different fields. Not only daily necessities of Tibetan people, but also Buddhist supplies in Tibetan temples. It's related to Tibetan life, culture, religion, medicine and so on. And the commodities sold are very practical, cultural, artistic, and ornamental. Furthermore, the big difference in the streets is its typical ethnic culture. According to the field survey, Tibetan's account for a considerable proportion of the sellers although there are some Han sellers from Chengdu, Chongqing, Zhejiang, Fujian, and other places. For example, more than 90% of Buddha statues sellers are Tibetans. Most of them come from Tibetan areas in Sichuan and they are familiar with their own ethnic culture, understanding Tibetan and Chinese, and knowing about the needs of Tibetan consumers (Fig. 13.2).

Therefore, the commodities as well as the decoration style of the store are full of Tibetan cultural elements. Compared with other streets, the distinctive ethnic traditional culture of Tibetan commodity streets has added rich content and diversified forms to Chengdu's commercial trade and regional culture, and it has become an important part of Chengdu's urban multiculturalism.

13.3.2 Typical Spatial Characteristics of Tibetan Commodity Streets

With the constantly influx of Tibetan population, the supply and demand of Tibetan commodity in the streets is continuously rising, the commercial atmosphere of the streets is becoming increasingly strong, the stores are expanding and relatively concentrated, and the types and quantity of commodities are increasing, which makes the economic center status of the streets more and more conspicuous. While the aggregation function of commercial economy increases, the scale effect of commodities consequently increased. It generates a more powerful economic attraction, and promotes the geospatial scope of Tibetan commodity streets to expand outward along the core areas of Ximianqiaoheng St., Wuhouciheng St. and Wuhouci East St. For example, it expands toward Gaoshengqiao, Guangfuqiao, Hongpailou and Taipingyuan areas along the South Fourth Section of the First Ring Road of Chengdu city. The geospatial characteristics of Tibetan commodity streets in the urban city are becoming more and more typical.

Consequently, the gathering of Tibetan commodities has brought about the concentration of Tibetan population living around here. Therefore, this area, which has always been spontaneously formed, relatively concentrated, and gradually expanded outward, is often called "Tibetan Commodity Wholesale Center", "Little Lhasa in Chengdu" and "Bakuo St. in Chengdu". This commodity street with strong geospatial characteristics are very rare in the first and second tier cities in China.

13.3.3 Obvious Localization of Tibetan Commodity Production

The scale effect of Tibetan commodity selling in Chengdu has brought about the change of Tibetan commodity producing center. The places with only one hour's driving distance to Chengdu city, like Chongzhou, Qionglai, Pidu, Xindu, Dujiangyan and Dayi, owns convenient road conditions, rapid information and advanced technology, sufficient human resources, low producing costs in workshops and land. It results in low producing costs of Tibetan commodities and then takes the lead in becoming the producing centers of Tibetan commodities.

For example, Ranyi Town of Qionglai City, the largest producing base of ethnic products in Western Sichuan, produces Hada, valance, Pulu, Nepalese door curtain, butter lamp, Buddha lamp, Buddha lamp oil, King Kong knot, prayer flags, paper Longda, Mani stone, Scripture cloth, printing and dyeing products which are all sold well in Tibet, Qinghai, Gansu, Xinjiang, Inner Mongolia, and other places. Some products are even sold to India, Nepal, Sikkim, Myanmar, and other countries and all of these are very popular among Tibetan consumers all over the world.

The long-term commercial trade cooperation between Tibetan commodity producers and sellers has not only generated mutual understanding and trust, but also further consolidated the trade cooperation relationship between them. Then, relying on the development of Tibetan commodity economy in Chengdu, the producing scale and field of Tibetan commodities in the surrounding areas of Chengdu city are also expanding. Therefore, the localization of the Tibetan commodities production is getting more and more obvious. The obvious localization trend of Tibetan commodity production further promotes the enhancement of the market scale and sales potential of Tibetan commodities in Wuhouci District of Chengdu.

13.3.4 Less Related and Dependent Between Tibetan Commodities and the Needs of the Residents

It is also learned from the field survey that the Tibetan commodities sold in the streets are mainly to meet the need of Tibetan people's daily life as well as the religious belief of Tibetan temples. The main consumer of commodities is still the Tibetan people all over the country while the correlation and dependence with the life and work needs of residents in Chengdu is almost zero.

13.4 Spatial Formation Mechanism of Tibetan Commodity Streets in Chengdu

13.4.1 Historical and Geographical Factors

As the provincial capital of Sichuan, Chengdu has a long history, pleasant climate, rich products, beautiful natural environment, convenient transportation, and talent gathering. It is not only the political, economic, and cultural center of Western China, but also an important gateway for Tibetan areas and the other places since ancient times, which plays a very important role as a link hub. First, since the Han Dynasty, Chengdu has had the trade tradition of "tea horse exchange" between Han and Tibet. Therefore, Wuhouci area in Chengdu has become a transit place for Tibetan caravans to enter the Central Plains in history. During the Ming and Qing Dynasties, it became an important starting point for Tibetan areas to enter the Central Plains and became the main place for chieftains and leaders from Tibet to stay.

After the founding of People's Republic of China in 1949, schools, government agency, banks, hospitals, and other institutions were set up around the streets, such as Southwest Minzu University, Chengdu office of the Government of Tibet Autonomous Region, Chengdu office of the Government of Ganzi Tibetan Autonomous Prefecture, Chengdu offices of other Tibetan counties and bureaus, the branch Tibetan hospital of Huaxi Hospital, and the branch bank of Lhasa People's Bank of China. It attracted many Tibetan people to study, receive medical treatment, work, and deal with affairs. Some chose to settle here. The settlement area for the Tibetan mobile population is gradually formed around this area in Chengdu city.

With the reform and opening-up of China, a series of favorable policies have been implemented to promote the continuous development of the market economy, the exchanges, and communications among all ethnic groups. The demand market in Tibetan areas has been expanding, the space and vitality of Tibetan economy have been continuously released.

In addition, Chengdu city's unique location and humanistic advantages, more relaxed market environment for selling Tibetan commodity and improved transportation and information network conditions make Tibetan commodity trade in

Chengdu more frequent and active. The increasing scale of Tibetan commodity further promotes the concentration, marketization, and specialization. It leads to the market agglomeration effect. Thereafter the core Tibetan commodity streets can finally be formed around Wuhouheng St., Wuhou EastSt., Ximianqiaohong St. It gradually replaces Ya'an city as the new center of Tibetan commodity production, sales, and distribution.

13.4.2 Population Agglomeration

Push–pull theory is one of the basic theories to study population mobility and urban–rural migration. The reason why people move into the city is that people can improve their living conditions through migration. Modern push–pull theory holds that in addition to higher income, the push–pull factors of migration also include better occupation, better living conditions, better educational opportunities for themselves and their children, and better social environment. Therefore, the demographic factor is also very important to the formation of urban commercial centers.

Since the 1990s, China's urbanization process has led to the cross regional mobility of population in a large scale. Since 2003, Chengdu has been approved by the State Council of China as a pilot city for the comprehensive reform of urban and rural areas. Through the reform of registered residence, social security, basic public services and rural property rights system, the integration of rural population and urban population in identity and welfare has been achieved. The implementation of a series of policies further promotes the inclusive development of Chengdu. In 2019, Chengdu ranked third in the cities with population inflow in whole China, which further shows its strong attraction to population.[4] Therefore, more employment opportunities, relatively high-quality education and medical conditions, better natural climate and environment and other factors release a strong pulling power which attracts more and more Tibetan population.

It is obvious to see that more and more Tibetan people move to Chengdu rather than other cities. Especially after 2000, many Tibetan people in Tibet, Ganzi and Aba Autonomous Prefecture began to move into Chengdu of their own will. At present, Tibetans are the most populous ethnic minorities in Chengdu. Due to the most concentrated distribution with the most distinctive cultural characteristics around Wuhouciheng St., the nearby residential community, such as Jinjiangyuan, Wuhouminyuan, Chaoyang Garden, Chengdu A, and the surrounding areas of Luoma Holiday Plaza, have also become popular choices for Tibetan people.

The cross regional and large-scale inflow of Tibetans into Chengdu also shows that the Tibetan population fully recognizes the inclusiveness of Chengdu city. According to the evaluation survey on the relationship between Tibet and Han, 74.6% of the Tibetan respondents think it is better and 21.5% think it is good. On the whole,

---

[4] Yeyan, Zouyue. Decoding the attraction of Chengdu. Chengdu Business Daily, July 23, 2020. https://e.chengdu.cn/html/2020-07/23/content_680265.htm.

the Tibetan floating population has a high evaluation of the relationship between Tibet and Han [1].[5] On the other hand, according to the statistical data of the sixth national census, the number of ethnic people in Chengdu has increased greatly. Among them, the Tibetan population in Chengdu totaled 32,332 in 2010, ranking first in Chengdu. In terms of distribution, Wuhou District has the largest Tibetan population compared with the other five main districts of Chengdu. with a total 15,212 Tibetans[6] (excluding the data of Tibetan temporary residence). A surge of immigrants brings about the agglomeration effect of population, and then brings the agglomeration of capital, information, technology, and other factors, which promotes the continuous development of Tibetan commodity economy and trade and endows Tibetan commodity streets with the new vitality of economic development.

13.4.3 Inclusive Development of Urban City

As a social system, urban city's inclusive development includes its opening, attracting the settlement of mobile population, and making them enjoy equal treatment with local people, rather than being excluded from urban life, work, and public affairs. All citizens, including the migrant population, have the same values and behavior patterns [2].[7] As for Chengdu City, it has been an immigrant city since ancient times. Tianfu Culture, which is characteristics of compatibility in thousands of years of historical evolution, has a great impact on Chengdu. It gives the city the characteristics of "compatibility", "learning" and "friendliness". Combined with its unique location, Chengdu has become the most important hub connecting the inland city and Tibetan areas. And the role cannot be replaced by other cities.

In this case, Chengdu Tibetan commodity street is formed and developed under the comprehensive influence of its nature, history, geography, economy, society, and culture. As a spatial existing form of commercial and economic activities, its economic development orientation, scale, speed, and level must be based on the development of Chengdu city [3]. Firstly, as a new first tier city in China, Chengdu is an important core city as well as trade logistics center in Western China, which has strong economic attraction and radiation to nearby provinces and cities. In 2021, Chengdu ranked 12th in the overall economic competitiveness of Chinese cities and first in the western region of China.[8] At the same time, Chengdu has an important

---

[5] Cunfang [1] Investigation Report on the Living Conditions of Tibetan Mobile Population in Wuhou District, Chengdu, Tibetan Studies of China, vol.2, pp. 15 (2012).

[6] The population of ethnic minorities in Chengdu is increasing–the fourth special information of the sixth national census, Chengdu Bureau of Statistics, http://www.cdstats.chengdu.gov.cn/htm/detail_27935.html, July 5, 2011.

[7] Jingxi [2] Inclusive development: the guiding choice of China's Urbanization–an analysis based on the principle of social system evolution, Social Science, vol.11, pp. 67 (2011).

[8] Pengfei, Ni, Ranking of comprehensive economic competitiveness of Chinese cities in 2021, The 19th report on China's urban competitiveness, Chinese Academy of Social Sciences, http://www.ttpaihang.com, Nov. 22, 2021.

transportation position in the western region with various modes of transportation, which drives the formation of a regional business center. Then commodities can be quickly transferred from the place of production to consumption market and ensure the market supply of the Western China in a timely and efficient way. With the great support of national policies and the boost of market economy, Chengdu economic development is getting vitality in a short time, and it provides a strong base for the formation of urban Tibetan commodity exchange and distribution center.

Secondly, as one of the "most influential cities in the Western China" and one of the "happiest cities in China", Chengdu, a famous historical and cultural city, has strong urban inclusiveness, which not only enables the Tibetan population to be fully respected, accepted and recognized, quickly adapt to the city's cultural life and economic development and love this city, but also provides a realistic possibility for the urban mobility of Tibetan population and the exchange of Tibetan commodities.

For those businessmen from other areas, all departments of Chengdu government have not blocked or excluded them. Instead, they provide social management services in different ways. For example, Chengdu Service Center for Ethnic business affairs, Chengdu Comprehensive Information Center for Ethnic people, "Association of Ethnic Traders",[9] established in Wuhou District of Chengdu with high-quality and convenient services in both Tibetan and Chinese which is helpful in reducing communication barriers. And the public squares in Ximianqiao St. and Wuhoucheng St. of Wuhou District are provided for Tibetan Guozhuang Dance Party which enhances the cultural adaptability of Tibetan people in the city and the continuation of Tibetan traditional culture [4, 5]. Also, many Chengdu residents would like to participate in Tibetan cultural activities, and it enhances mutual understanding. Meanwhile, employment and entrepreneurship skills training are carried out to help Tibetan businessman and other mobile population to improve the business skill and employment ability. Therefore, from the Tibetan commodity production area to the Tibetan commodity sales area, and then to the Tibetan commodity consumption area, the efficient and smooth connection of economic activities in production, sales and consumption has been realized, which fully shows the strong openness and inclusiveness of urban socio-economic and cultural development.

13.5 Conclusions

Chengdu city has a complete ethnic composition and a large ethnic population. The equality, unity, harmony, and symbiosis of ethnic relations have brought about social stability in Chengdu and even the whole Western China, the improvement of urban commodity economy and the development of urban multiculturalism. It also provides some experience for urban development in this world. Due to its complex

---

[9] Chengdu explores the establishment of a long-term model of embedded demonstration communities among ethnic groups, Chengdu Municipal People's government, http://www.chengdu.gov.cn/chengdu/home/2018-01/17/content_a7c91359fa424dc2bd2c5feac2f8b454.shtml, Jan. 17, 2018.

ethnic composition, diversified ethnic culture and "mixed and concentrated" living pattern, and unique Tibetan commodity selling, Wuhou Tibetan commodity streets in Chengdu becomes a model for the study on multi-ethnic community governance. Today, this equal, United, mutual aid and harmonious symbiotic ethnic commercial streets has become an urban ethnic exchange and integration center, an ethnic economic center, and an urban multicultural development center.

At the same time, the development process and current situation of Tibetan commodity streets in Chengdu also reflects the city's multi-cultural connotation and the new characteristics of urban socio-economy. It makes the city more dynamic, socially inclusive, and culturally attractive, which complement each other and then promote the virtuous cycle of urban socio-economic development.

References

1. Cunfang D (2012) Investigation report on the living conditions of tibetan mobile population in Wuhou district, Chengdu. Tibet Stud China 2:13–15
2. Jiangxi H (2011) Inclusive development: the orientational choice of urbanization in China-analysis based on the evolutionary principles of social systems. Soc Sci Res 11:64–67
3. Mingdou Z (2015) On comprehensive measurement and spatial effect of urbanization level. Res Financ Econ Issues 10:138–142
4. Renzhen L (2012) Relationship between han and tibetan ethnic groups from the Tibetan mobile population in the inland-take tibetan mobile population in chengdu as an example. Tibetan Studies of China 2:11–13
5. Xueshu X (2015) On the management of tibetan mobile population services: taking Chengdu, Sichuan province as an example. J Southwest Minzu Univ (HumIties Soc Sci Ed) 6:33–37

Chapter 14
Rescaling of Chinese Urban Space: From the Perspective of Spatial Politics

Xiaoxi Liu and Qianning Li

14.1 Introduction

As the argument of Micheal [1] in the Eye of Power, a complete history (also as a history of power) needs to be written by space, including major strategies from geo-politics to small strategies at the residential level, and various economic and political measures in it. Since 1970s, capitalist urban space production and the political function of space have also become main topics of Marxist theorists. Applying this theoretical perspective into the changes in China's urban governance model, space has also become an important governance strategy. Urban governance model in China has gradually transformed since 1949, and it can be generally divided into two stages. The first stage is dominated by the Danwei system and supplemented by the community system during the planned economy period; the second stage emerged after the Chinese economic reform in 1978, with the community system playing an increasingly important role as well as the disintegration of the Danwei system. The form of space in China is flexible and changeable, and its internal structure is an important perspective for understanding the operation of the urban governance [2]. Currently, the community system has become the main organizational form of urban governance, especially in the grassroots level. There is a shared similarity: relative closedness in space, which presented in both Danwei and community. These spaces are defined by geographical boundaries or administrative boundaries, or both of the two. This paper attempts to discuss the relationship between space and system from the perspective of spatial politics, and tries to understand the operation of urban governance by combing the division method and internal structure of urban space.

X. Liu
Department of Political Science, Tsinghua University, Beijing, China

Q. Li (✉)
School of Government, Nanjing University, Nanjing, China
e-mail: dg1906007@smail.nju.edu.cn

14.2 Literature Review and Theoretical Framework

14.2.1 The Political Characteristic of Space

In terms of the theoretical debate of space and politics, the research work of Henri [3] presents representative and fundamental significance. Lefebvre brings the perspective of space and geography into Marxism, focusing on "production of space" rather than "production in space". Edward [4] divides space into three main types, including natural physical space, psychological space, and space as social production. The state, as a typical representative of social production space, can in turn promote the spatial reproduction in a specific society. The argument of David [5] illustrates that each social form constructs objective concepts of space and time in order to meet the needs and goals of material and social reproduction. Capital plays an important role in promoting space production, and the fair distribution of space resources is also of great significance. Such arguments follow the same logic, concentrating on the function of space production while failing to break away from the perspective of urban cities' social and economic functions.

Neil [6] links the form of urban governance with national spatial control strategies, and provides an institutional perspective for the discussion of spatial theory. Through the research on the experience of urban governance in Western Europe, Brenner found that the European urban governance model has under profound changes: at first it was embodied in the management-welfarism where the nation-state was mainly responsible for providing welfare, and then due to the decentralization of urban governance power from the national level to city level, local governments mainly assumes the responsibility of urban governance and actively promotes urban development. The shift of the attention from economic and social functions to urban governance functions highlights that there is another feature about space: the political functions of space.

While defining the boundaries of governance, space also shapes the basic content and complexity of governance activities [7]. On the one hand, urban governance affect the production of social relations through the transformation and adjustment of physical space. Therefore, governance based on the space first means a kind of "Cartography", which converts complex physical spaces into abstract spaces that can be analysed and controlled [8]. On the other hand, adjusting internal social relations of the space can also be used to reshape the physical space [2].

Since space is of political and strategic importance, what are the characteristics of the political system formed in a specific space? Does the scale of the space interact with the effectiveness of the system? The answers to these questions require the introduction of relevant theories of scale politics, and further analysis from the perspective of the formation and changes of Chinese urban space is also needed.

14.2.2 The Scale of Dividing Space

Modern cities have generally established sophisticated spatial systems and spatial orders, and the production of urban space itself has been largely determined by political policies from the government. The urban spatial form reflects strong characteristics in urban planning, and the space with clear boundaries also reflects its clear professional functions [9]. Governing societies through urban planning policies has become one of the important means to control and regulate urban social life, and the changes in spatial forms have also become an important manifestation of the reform of urban governance practices. Another branch of literature concentrating on debates of space comes from the research in the field of political geography. The spatial scope is limited by the division of geographical boundaries, and the politics of scale has become an important perspective to understand the changes in specific urban governance system.

As one of the core concepts of geographical research, "scale" is generally used to illustrate spatial phenomena, especially in the field of natural physical space [10]. With its usage extending into the field of political geography research, the concept of scale starts to contain constructive meaning. In this sense, scale is one of the important factors affecting space, covering meanings such as size, proportion, and level. Whether the discussion is relating to political geographic space or social relational space, most of the spatial production strategies are based on scale [11]. The space unit is not static, but can be adjusted and reorganized according to the needs of the function, and the space can be parsed and refined [12].

In terms of "politics of scale", it was first proposed by Neil [13]. Through analysing the scale application strategy of the homeless, he pointed out that scale is the framework of various social movements and struggles. Brenner [14] expanded the sense of scale and used it to discuss regional restructuring in the process of globalization, regarding scale as an important governance tool. Scale can be regarded as an operating mechanism of power in specific political systems.

Sallie [15] proposed another concept: "rescaling", and extended it to the activities in social reproduction. In the operation of the political system, rescaling not only means the dissolution of the old mechanism, but also the emerge of a new mechanism. The reason for this change is the change in power relations. The change of power between different scales forms a process of new scale system and political strategy [16].

14.2.3 Systems within the Space

When recognizing China's administrative divisions under the perspective of scale, the existing research on scale reconstruction mostly based on the empirical induction of local government practices. Some domestic scholars have conducted research of scale reconstruction through discussing the adjustment of administrative divisions in

China, such as "provincial direct management" or "removing counties into districts", exploring the phenomenon and consequences of the reform to place counties and county-level cities directly under the jurisdiction of provincial governments or transform counties into urban districts [17–19]. There are also other relevant researches focusing on urban agglomerations which transcend urban boundaries, including typical cases like the Yangtze River Delta, Pearl River Delta, Beijing-Tianjin-Hebei and other regions [20], or analysing new urban districts, development zones, and new districts of national-level that established by adjustment of administrative divisions [21].

Independent cities under the administrative hierarchy are gradually merged into the category of city-regions for discussion. The space defined by the administrative scope is not only a natural space in the sense of geography, but also represents an institutional space formed by administrative policies [22]. Most of the existing studies have focused on the discussion of spatial integration across administrative divisions, focusing on the adjustment of rigid scales, but its change flexibility is small and has long-term stability. With political forces intervening in system reconstruction and governance model reconstruction, the reconstruction of the regional space is one of the basic premises of the reconstruction of the scale. It is not only the change of natural space, but also the change of the regional characteristics of the administrative space. It can also be regarded as a reform of the reconstruction of the scale [23].

The geographical area covered by the city blocks can also be used as a microanalysis object showing spatial politics. For this kind of political system space, related planning activities are strategic behaviors with spatial politics. In urban administrative spaces, there are often tensions between different spatial units, which can be regarded as a special type of scale [24]. The reshaping of the political system space within the existing administrative divisions is reflected in the government's re-political integration of regional space on the basis of policy planning to mobilize spatial production. The districts and city blocks within the city can also act as an object to be analysed from spatial politics [22, 23].

14.2.4 Limitations of Existing Literature

After reviewing relevant literature, there are two academic branches in terms of politics of space and of scale, illustrating urban space from different perspectives. However, these theories about urban space presents mechanism of power in political space also present similarity. Based on the productive, political and instrumental nature of space, the production of political systems can be discussed together with the issue of scale reconstruction. In addition, administrative systems operating by different departments in grassroots level of urban Chinese cities are organized on the basis of urban space as well. However, there is not enough concentration in previous literature on the relationship between specific forms of urban space and the administrative systems within them.

The social-economic space across administrative divisions will inevitably involve the change and reorganization of the governance system in the governance process. The original governance subject and the organizational structure of the governance objects are faced with new changes, which also promotes the formation of a new political system space. Furthermore, what are the regular forms of urban space and how do the administrative systems operate in specific urban space unit in different periods of urban development in China? This research try to integrate the above discussed branches of theories and explore the relationship between space and political system.

14.3 The Formation and Rescaling of Urban Space in China

14.3.1 The Space of Danwei: Scaling Under the Occupational Boundary

"Danwei" emerged before the establishment of the People's Republic of China. In the process of urban construction, Danwei have been established in the form of systems as spaces for urban residents to work and live, and have become the main carrier of Chinese society [25]. The predecessor of the Danwei system originated from the supply system in during period of "Revolutionary base area". Specifically, it refers to the conditions of supplying individuals with the most basic daily necessities on the principle of "general average" because of scarcity of material materials. In the later stage of development, the items supplied have increased to various aspects such as clothing, food, housing, transportation, etc. [26].

Andrew [27] regards Danwei as a general industrial personnel management mechanism in socialist countries to some extent. In 1948, some provinces in Northeast China took the lead in liberating the cities of Harbin, Changchun, Shenyang and other provincial capitals, which was the beginning of the establishment of a universal system for Danwei [28]. Since resources are highly concentrated and relied on the unified deployment of the country, various institutions and companies are organized in accordance with the principles of unified planning, centralized management, and overall mobilization. In Chinese academia, this social structural feature is also known as "Macromanagement" [25]. At this stage, Danwei serves as a bridge between the state and the individual, and has become synonymous with society. The concept of Danwei overlaps with the concept of society infinitely [29].

In urban China, Danwei system played an essential role in the planned economy period (1949–1978), especially in the activities of political mobilization and public goods' supervision [27]. Typical Danwei organisations are mostly state-owned enterprises or public institutions, which are invested by the central government instead of relying on the local government's fiscal and taxation system (Bray, [30]. Danwei is not only a kind of organisation in producing goods, but also operating as an integration mechanism in provision of various services to members that included in it [28].

Danwei space then turned into an institutional mechanism in urban China, which was bounded by closed high walls reflects the closeness of the space, and is similar to the family courtyard under the traditional Chinese patriarchal system. This form of space presents as a modern technology that can be used to maintain the social order of collectivism [31]. The strong organizational characteristics of Danwei and the relative closedness of space have shaped the production system and governance system of socialist China based on public ownership. The overlap of Danwei organization which used to social production and the space formed based on it, as well as the reduction of social mobility, enable the state to regulate the management of organized individuals [31].

The Danwei system is established under the premise that the government conducts overall management of enterprises and institutions. In this management system, organization and mobilization based on occupation and the provision of benefits have become important system features. The spatial structure of Danwei reflects the characteristics of the "combination of work and residence". Thus the space of Danwei contains two main functions: workplace and residential space [2]. The discipline of the Danwei for its employees is not only reflected in the work field, but also penetrates deeply into the residential places of private life. Residential areas of Danwei integrate workplace, life service functions, employees and their families, establishing a social network within the community [32]. Danwei is not only a workplace but also a support for collective life. Under this spatial arrangement, the spirit of collectivism is shaped, because workers will not leave the organization life after their labor [31].

It should be noted that the space shaped by Danwei is "relatively closed". Generally speaking, in addition to work facilities, Danwei are often responsible for providing residential facilities, canteens, shops, bathrooms and other living facilities as well as welfare services such as education, medical care, and entertainment. However, due to differences in organizational scale and resource control levels between different types of Danwei, the supply coverage of these facilities is also different. Some can maintain complete internal supply, while others need to rely on external resources to varying degree, such as cooperating with regional commercial center for auxiliary supply [33].

During the period of the planned economy, Danwei undertook specific governance units based on the workplace, and the occupational boundary naturally became the spatial boundary and institutional boundary under the organizational form of the Danwei system.

14.3.2 The Space of Community: Rescaling Under the Residential Boundary

The community system also emerged in the planned economy period (1949–1978), acting as an auxiliary system in integrating urban society in parallel with Danwei. Since "the Regulations of the People's Republic of China on the Organization of

Urban Resident Committees" issued in 1954 (which was replaced by "the Organic Law of the People's Republic of China Urban Residents Committee" in 1990), the Sub-district Office and the Urban Resident Committee had been operating like administrative function. There are two major integration mechanisms formed in urban society: "State-Danwei-Individual" and "State-Community-Individual" [34]. However, the function of community space played a very limited role till the institutionalized Danwei space started to fade out [28].

After the Third Plenary Session of the 11th CPC Central Committee in 1978, great changes have been promoted in various domestic industries at the structural and institutional levels. The rapid economic development has also promoted the acceleration of the urbanization process, a large number of floating populations have appeared, and the development trend of occupational diversity has also caused the internal members of the Danwei system to flow out, and social members are no longer stable under a fixed unit [33]. As the central government reduces investment in Danwei and local governments obtain large amounts of fiscal revenue through land sales, the internal structure of government organizations has also changed accordingly. In order to strengthen the management of labour force outside the Danwei system, the administrative power of sub-district office and urban residents' committee began to rise. The power of local governments to adjust urban planning has increased. In the new urban planning system, the most basic spatial unit falls in the community.

In 1984, the "Decision of the Central Committee of the Communist Party of China on Economic System Reform" was regarded as the beginning of urban reform. One of the important goals of the reform was to reduce the welfare function of Danwei and allow local governments to provide basic community services to the urban population. In the subsequent process, community-based activities are no longer limited to services and welfare, extending to education, culture, sports, health, comprehensive governance, grassroots democracy, etc. This also means that the community has replaced Danwei as a new basic unit of urban governance. In 1987, the Ministry of Civil Affairs advocated the development of community services for specific targets in cities, and the concept of "community" entered the management process of the Chinese government for the first time. With the popularization and deepening of community services, community service targets have gradually expanded to all community residents, and the projects involved have become increasingly extensive and diversified.

Since the market-oriented reforms, in terms of urban spatial distribution, business districts, industrial districts and living communities in the city have been showing an increasingly obvious spatial separation, especially between the occupational space and urban residents' living space. The privatization of housing property rights, the complexity of population composition, and the externalization of residential space have made the composition of residents in the same residential area more complicated. The original occupation-based living space is gradually changing to a social space based on daily life [30]. The relatively fixed correspondence between population and space has gradually transformed into the flow of population in different spaces. The object of governance activities is no longer a relatively fixed, organized population, but is transformed into the governance of space. In other words, governing urban society is now faced with challenges of addressing issues in specific spaces [2].

Community space can be regarded as a comprehensive system in the physical sense, containing a series of factors like roads, green parks, education service, commercial service and other functions [34, 35]. A large number of communities have developed functionally-rich physical spaces and built public venues covering multiple functions to provide residents with physical activity areas required for daily activities and neighbourhood communication. In addition, Community space can play a role as a field of power interaction, and it has an interactive and mutually constructive relationship with its internal multiple actors. The actors in the space and the space have mutual influence. Governance activities conduct people's behavior, but there are also counter-conducts against governance [36].

The development of the market economy has increased the flexibility of individual's career choices, resulting in the mode of organization and mobilization from a closed working space is not able to cover the entire urban space anymore. Unlike the Danwei system, which has obvious professional attributes, the space scaled by community system is divided into administrative orders and residential areas as the main scope. As community has become the basic unit of urban governance, space of Danwei has gradually transformed into community space. In fact, the division principle of urban space and the supporting conditions behind it have undergone substantial changes.

14.4 Discussions

14.4.1 A Comparative Framework: Spatial Practice After Rescaling

On the basis of the previous discussion, China's urban space has been rescaled from the space of Danwei to the space of community, and changes in spatial form also derive differences in political system's operation. The above two different spatial scales are important institutional tools for the application of urban governance activities in grassroots society.

When individual behavior is placed in a complex urban space, its cognition, preference and selection process are mostly restricted by the space. On the one hand, a relatively fixed community space is the basis for residents to carry out individual activities or organized actions. In addition to physical space, factors such as social structure, institutional arrangements, and social relations are also included; on the other hand, there are cooperation, exchange, competition and conflict with others when the actors pursue preference choices of their own benefits. In this process, the actors influence and reshape the rule of power in the community space [37]. The systems and strategies deployed in the space, together with the conflicts and contradictions in it, both constitute space practice. In addition, the interaction between contradictions and systems has also become a motivation for the development of social and political practice.

Space practice includes not only the construction of political systems, but also the conflicts between different actors in the space. Because space is political, strategic, and full of ideological expressions, space is not only a tool of production and consumption, but also a tool of domination and resistance; is not only the basis of any public form, but also the basis of any power operation [38]. Compared with Danwei that mainly perform organizational functions, the community area presents more interaction. Community space can play a role as a field of power interaction, and it has an interactive and mutually constructive relationship with its internal multiple actors.

As the community system has become the focus of urban development, combined with the increase in resources held by local governments, urban planning has become an important driving force affecting the shaping of community space. From the perspective of community space production, housing commercialization has triggered changes in the power structure of communities, and rights protection awareness of individual residents and social organizations also emerged. The rights protection and resistance activities of owners are behaviors that appear after the reshaping of urban social spaces. This change shows that the community is not only a place of residence in the physical sense, but also a public place where various actors carry out power operation and interaction. The community residents who live together in the same public domain tend to have common problems and demands in this definable space.

In addition to the space production with the government as the main body under the power logic and the space production with the market as the main body under the capital logic, the participation of residents is also a kind of institutional behaviour that shapes space. The urban public space formed by public communication is an important place for residents in a certain area to carry out public life. It can play the function of balancing the private sphere and help reshape personal identity. Space practice is closely connected with daily life and can encourage individuals to maintain a specific spatial identity and produce their own space. Different actors in a specific space sometimes use "weapons of the weak" to shape their own living space. The spatial practice of urban communities is not only embodied in the contradictions and conflicts of residents in daily life, but also composed of the interaction between residents and the government. Many space conflicts occur because the space practices promoted by the government from outside violated the "social space boundary" that formed spontaneously by residents [34]. In this way, the construction of the relationship between the government and residents in the community space is more embodied in the game process of "conduct" by technological governance and "counter-conduct" by residents through spontaneous actions. This undoubtedly increases the complexity of urban space governance.

14.4.2 A Further Step from the Community: The Space of Grid

The change of the urban system from the Danwei system to the community system also reflects the transition from "macromanagement" to "micromanagement" [25]. Innovations in urban governance from the perspective of space has been undertaking the process of updates and transformations. The following stage of rescaling urban space is dominated by the space of grid. Owning to the formation of grid is on the basis of community, resulting in the necessity of comparing these different space forms and exploring similar mechanisms in rescaling of urban space. The space of grid can be understood as an update of the community space, which is evolved from the Grid-based policy that generated in 2004, aiming to address important issues such as the provision of public services. Dongcheng District of Beijing was selected as the first pilot for exploring this practice of urban space rescaling process. GPS, GIS and a series of geographical technologies were used in defining grid unit, and finally the 126 communities of Dongcheng District were divided into 1593 grids [39]. The practice of Grid-based policy was initially carried out at the local government level, aiming at developing digital cities, relying on a number of network technologies and geocoding technologies, and it is an example of the systematic application of information technology to the practice of local governments. With the expansion of the expansion scope of this policy, grid has gradually become a generally applicable governance technical means in a comprehensive range. After nearly 20 years of development, the setting of grid space and the construction of governance structure under grid have been preliminarily completed nationwide.

Foucault [40] illustrates that the governance of population is not only related to collective and large-scale phenomena, but also contains in-depth management of population in details. The governance of population can also be achieved through space technology. In academic research on Chinese urban space, some scholars compare the Grid policy with the "Fengqiao Experience", regarding them as governance techniques for dealing with social contradictions in a specific space. Combining with [41] discussion of community corrections, dispute settlement and judicial assistance from the perspective of comprehensive governance of grassroots society, it is easy find that the initial application of grid technology is targeted on specific groups, which purpose is to maintain social stability. In this circumstance, the political and legal system undertakes the main responsibility to carry out social correction activities. The initial purpose of implementing the grid policy is to ensure the stability of the grassroots and resolve social conflicts. In Dongcheng District, Beijing, grid is generated to solve the weak links and main difficulties in stability, focusing on limited groups of people that may harm social security [42]. A refined management platform based on modern information technology is also built to perform functions such as collecting and reporting major issues, assisting in public security prevention and control, and mediating conflicts and disputes. It could be concluded that the main responsibility of the grid in the early stage is to maintain governance function of social stability.

The application scope of grids has continued to expand, acting as a governance technology on the basis of specific governance goals [43]. A number of public service and management tasks are implemented at the grid level, the function of grid space has been shifting from focusing on management to both on governance and services. In addition, the objective of grid governance also expands from specific group to general dwellers geographically, including all population, places, objects, and organizations in the grid space [44]. The expansion of focused groups indicates that the main function of grid governance has been transformed from regulatory governance to providing social services [45]. The basic tasks of urban governance at grassroots level are no longer limited to maintaining stability and reducing crime rates, which means the governance function of grid has gradually begun to shift from stability-oriented control to the provision of service functions. The application of governance technology is not only for a certain part of population who may bring danger and create contradictions, but develops into all population groups which are contained in the specific space.

As an important innovative practice of urban governance, Grid-based Policy has been promoted from specific pilot cities' policy experience into a nationwide institutional construction process. After over 17 years' development, dividing urban space by grid has developed into a national institutional measure, and "Governance in the Grid" also becomes an important govermentality of local governments. After years of piloting and promotion, the Fourth Plenary Session of the 19th Central Committee of the Chinese Communist Party formally proposed to "Improve the service mechanisms of the community governance as well as the grid governance." The importance of implementing grid governance has been deeply stressed by the central government. Compared with the vertical pressure of the provincial government, the policy initiative at the central level plays a more obvious role in promoting; the neighboring pressure generated by the policy adoption of other cities in the same province can also effectively promote the adoption of the grid-based policy.[1] Grid governance has become an innovative measure of urban governance practice and is regarded as the guarantee of precise and refined service by the government. The division of the grid is based on the basic structure of the community, but it can also be used as a reorganization of the internal space of the community.

The boundary of the grid system in urban governance is also defined by the boundary of the grid space, which is embodied as a space area with no gaps, no overlap, and no omissions. The grid is to refine the internal space of the community, using buildings instead of residential areas as the distinguishing unit, reducing the coverage of the first-level space unit. The division of the grid is based on the basic structure of the community, but it can also be used as a reorganization of the internal space of the community. The basic governance unit of the city is further transformed from the community to a smaller spatial form in the geographical sense.

Since the grid space represents as the latest space unit of urban China, administrative system also changed with spatial transformation, as the local governments' departments have been assigned important responsibility [39]. Each grid is supposed

---

[1] These arguments come from another forthcoming paper by the authors.

to be equipped with one or more grid administrators[2], in addition, where a grid with multiple grid administrators can be equipped with a grid manager. Except for the grid manager is delegated by superior department, other grid operators are generally composed of volunteers, social workers. In the grid space, the issued grid administrators act an essential role, including discover the problems in regular governance and report these problems through local E-governance website to the responsible departments.[3] In addition to the necessary gird administrators, there are a large number of volunteers assisting the implementation of grid governance, such as "building administrator" in each building of specific grid and "cell administrator" in each building.

14.5 Conclusions

The form of space and its internal practical activities are important tools that influence the practice of urban governance. They can be used instrumentally in specific governance activities through the production of political systems and as a medium for power to play a role. China's urban space has a unique early form and has under great changes. The Danwei system compound with the professional identity as the space boundary is enclosed by closed high walls and becomes the original basic scale of Chinese urban space. Once the spatial scale is established, it relatively stable and can adapt to the political system operating in the space.

However, the spatial scale is not static. Under the influence of national policies and market mechanisms, the inherent spatial form with Danwei as the main body has changed and gradually evolved into a new, more fluid open space. The original urban governance unit was forced to adjust, and the occupation-based spatial structure also changed. The boundary of work could no longer cover the governance needs of a large urban population. The basic unit of urban governance needed to seek a spatial form which can present more adaptability to the high mobility of population. As a consequence, the community system has once again delineated the city's space in the form of residential boundaries. After the disintegration of the Danwei system, the community space has further become a basic governance unit that performs important functions, becoming another urban space and a new political scale determined by the logic of residence. In the process of urban development, the spatial transformation from "Danwei yard" to "residential community" reflects the reshaping of geographic scope and institutional boundaries. The actors in the community space make the connotation of the space practice richer than that of the Danwei space.

---

[2] In different cities, there are different specific titles for this position. Some are called "grid members", "grid information officers" or "grid supervisors" etc., their daily work content is similar in nature.

[3] Some contents of this part is obtained from the authors' observations and investigations during fieldwork in Shandong Province, Aug. 2020.

The basic governance unit of the city is further transformed from the community to a smaller spatial form in the geographical sense. The institutionalized operation of grid governance has had an impact on the organizational structure and operation process of urban governance. It can be regarded as another process of rescaling of urban space, redefining the community space by adjusting the spatial scale and using information technology to carry out a new round of spatial positioning for urban management. Thus it could be convinced that space can carry out institutional production, and the process of rescaling a space will also affect the reshaping of existing political systems. In urban governance which is faced with an increasing need of refining its activities, the spatial unit of an appropriate scale is of great importance to ensure the effectiveness of urban governance activities.

As discussed above, a main conclusion drawn in this paper is that the rescaling of urban space is not just a reorganization in a geographical sense, but also reflects another important meaning: the government's adjustment of power relations in a specific space. The Danwei space reflects more occupational basis because it undertakes the resource allocation of the central government, and the local government actually has a limited influence on individuals and organizations inside this space. Also, the working space and living space of Danwei members are inseparable, which become a double constraint on the behavior of these members: individuals can only obtain services provided by Danwei, so there is rarely anti-system activities in this period, otherwise they will no longer be able to obtain benefit from Danwei anymore.

This also explains why there are more activities of multi-actors in the community space: the separation of occupational space and residential space no longer restricts individuals from only obtaining benefits and services in one organization. Restrictions on meeting their own needs have been lowered, which has become one of the reasons urging residents to take actions. At the same time, the government's encouragement of community autonomy has also caused changes in the distribution structure of power in the community space. Different from the almost one-way and top-down power mechanism in Danwei, the power structure in the community reflects more genealogical characteristics. Due to the diversification of residents' demands, the government's public service provision also tends to experience market-oriented and selective changes.

Finally, this paper also briefly mentioned the power mechanism in the grid space. Differing from the community, the grid is faced with a "Seamless Government" that promoted by Russell [46]. In addition, staff such as grid administrators take a representative role for residents to a certain extent, in order to report issues and make a request to the relevant government departments, which is undoubtedly a restriction on the power diffused in the community. The government's response to the grid space has increased, but at the same time, changes in the political system have also led to a large part of the power being reclaimed by the government. This is also an aspect in which grid space is different from community space and Danwei space. This also echoes the Chinese political system's emphasis on the function of stability maintenance. After twice rescaling of urban space, the resulting structure is the strengthening of both service provision and public order, which may eventually lead to a certain degree of restriction on residents' own space actions.

References

1. Foucault M (1985) The eye of power. In Colin Gordon. Power/Knowledge: Selected Interviews and Other Writings 1972–1977 (trans: Gordon C et al). Pantheon Books, New York.
2. Li W (2018) Unitization of space: political party mobilization and space governance in urban grassroots governance. Marx Rity (6) (in Chinese).
3. Lefebvre H (2015) Space and politics (trans: Li C). Shanghai People's Publishing House, Shanghai (in Chinese).
4. Soja EW (2011) The spatiality of social life: a theoretical reconstruction towards transformation (trans: Xie L, Lu Z). Social Relations and Spatial Structure. Beijing Normal University Press, Beijing (in Chinese).
5. Harvey D (2002) Spaces of hopes. Edinburgh University Press, Edinburgh
6. Brenner N (2004) Urban governance and the production of new state spaces in western Europe, 1960–2000. Rev Int Polit Econ 11(3):447–488
7. Chen X, Yang X (2013) Space, urbanization and governance reform. Explor Contend (11) (in Chinese).
8. Du Y (2017) Cartography: a new perspective on national governance research. Sociol Res (5) (in Chinese)
9. Chen Y (2012) Logic of urban China. Life·Reading·Xinzhi Sanlian Bookstore (in Chinese).
10. Croval P (2007) History of geographical thought (trans: Zheng S, Liu D). Peking University Press, Beijing (in Chinese).
11. Wang R (2020) Understanding spatial politics: a preliminary analysis framework. J Gansu Adm Inst (4) (in Chinese).
12. Foucault M (2012) Discipline and punishment (trans: Liu B, Yang Y). Life·Reading·Xinzhi Sanlian Bookstore (in Chinese).
13. Smith N (1993) Homeless/global: scaling places. In: Bird J et.al Mapping the futures. Routledge, London.
14. Brenner N (1998) Global cities, glocal states: global city formation and state territorial restructuring in contemporary europe. Rev Int Polit Econ 5(1):1–37
15. Marston SA (2000) The social construction of scale. Prog Hum Geogr (2).
16. Shen J (2007) Scale, state and the city: urban transformation in post reform China. Habitat Int, 3–4.
17. Luo X et al (2010) Incomplete reterritorialization and reorganization of metropolitan administrative divisions: a case study of the districting of Jiangning county in Nanjing city. Geogr Res (10) (in Chinese).
18. Wang Y (2004) Research on the reform of the hierarchical system of counties and cities directly managed by provinces in China and local administrative divisions. Hum Geogr (6) (in Chinese).
19. Zhang J et al (2016) The negotiation game and strategy features between the upper and lower levels in the adjustment of administrative divisions: taking SS town as an example. Sociol Res (3) (in Chinese).
20. Yang L, Mi P (2020) Why urban agglomerations become national governance units. Adm Forum (1) (in Chinese).
21. Chao H, Li G (2020) Governance rescaling of national-level new areas and evaluation of their economic effects. Geogr Res 39(3):495–507 (in Chinese)
22. Chao H et al (2015) Spatial production strategies in national strategic regions from the perspective of scale reconstruction: based on the discussion of national new districts. Econ Geogr (5) (in Chinese).
23. Geng F (2020) On the logic of the political philosophy of space: taking the publicity of urban streets as an example. Acad Exch (1) (in Chinese).
24. Ma LJC (2005) Urban administrative restructuring, changing scale relations and local economic development in China. Polit Geogr (4).
25. Qu et al (2009) From "Macromanagement" to "Micromanagement": a sociological analysis based on China's 30 years of reform experience. Chin Soc Sci (6) (in Chinese).

26. Lu F (2003) The origin and formation of China's unit system. In: The Institute of Sociology, Chinese Academy of Social Sciences: Chinese Sociology, vol 2. Shanghai People's Publishing House, Shanghai (in Chinese).
27. Walder AG (1996) Communist neo-traditioanlism: work and authority in Chinese industry. Oxford University Press, Oxford
28. Tian Y, Lv F (2014) The changes of "Danwei Community" and urban community reconstruction. Central Compilation and Translation Publishing House, Beijing (in Chinese).
29. Liu S, Wang C (2016) Organizational sociology analysis of grid social management. J Theory Guid (3) (in Chinese).
30. Chai Y et al (2014) Spatial behavior and behavioral space. Southeast University Press, Nanjing (in Chinese)
31. Bray D (2005) Social space and governance in urban China: the Danwei system from origins to reform. Stanford University Press, California
32. Chai Y (1996) Danwei-based Chinese cities' internal life space structure: a case study of Lanzhou city. Geogr Res 1:30–38 (in Chinese)
33. Chai Y (2000) Urban space. Science Press, Beijing (in Chinese).
34. Li L (2019) The fragmentation of public service supply in the changes of urban grassroots social management system: based on the analysis paradigm of historical institutionalism. Adm Forum (4) (in Chinese).
35. He X (2019) Types of spatial structure and neighborhood relationship: a spatial perspective of urban community integration. Society (2) (in Chinese).
36. McNicoll G (2007) Michel foucault: security, territory, population: lectures at the college de France, 1977–78. Popul Dev Rev (4).
37. Zheng X, Liu Z (2018) Review of research on the power order of urban communities in my country in recent years (2011–2016). J Shanghai Adm Inst (5) (in Chinese).
38. Bao Y (2002) Modernity and the production of space. Shanghai Education Press, Shanghai (in Chinese).
39. Chen P (2006) Gridization: a new model of urban management. Peking University Press, Beijing (in Chinese)
40. Foucault M (2008) Governmentality (trans: Zhao X). In: Feng G Selected Reading of Basic Sociological Literature. Zhejiang University Press, Hangzhou. (in Chinese).
41. Wang S (2018) Fengqiao experience: the practice of grass-roots social governance. Law Press, Beijing (in Chinese)
42. Zhou L (2013) Grid management: a new exploration of grassroots stability maintenance in China. Acad J Zhongzhou 198(6):83–85 (in Chinese)
43. Tian Y (2012) The orientation and promise of urban grid governance. Study & Explor 199(2):28–32 (in Chinese)
44. Song G (2012) Griding: griding services and formatting management. Beijing Daily (in Chinese).
45. Ye J, Wu X (2019) Grid linkage governance: practical innovation of grass-roots social governance in the new Era. Theory Mon 454(10):137–145 (in Chinese)
46. Linden RM (2014) Seamless government: a guide to public sector reengineering (trans: Wang D et al). Renmin University Press, Beijing (in Chinese).
47. Gregory D, Urry J (2011) Social relations and spatial structure (trans: Xie L, Lu Z). Beijing Normal University Press, Beijing. (in Chinese).
48. Li A Action in space: the adjustment of Dashilan courtyard space and social relations. Int J Soc Sci (Chinese Edition) 36(1) (in Chinese).
49. Tierney TF, Foucault M (2008) Security, territory, population: lectures at the Collège De France, 1977-78. In: Senellart M (trans: Graham B). Palgrave Macmillan, London.
50. Wang M et al (2001) Nietzsche's ghost: Nietzsche in the western postmodern context. Social Sciences Archives (in Chinese).

Chapter 15
Global Cities and Business Internationalization: Towards a New Interdisciplinary Research Agenda

Abdelhamid Benhmade, Philippe Régnier, and Martine Spence

15.1 Introduction

Globalization seems to accelerate a kind of economic space dispersion. However, the concentration of business activities continues to rise in so-called global cities [1]. In the context of a new international division of labor, these cities are the only ones able to combine global functions of business coordination, clustering and networking [2]. Internationalizing firms are facing more and more uncertain markets due to the constant reorganization of global value chains. Consequently, they need to tab on new types of external resources, which are commonly designated in the literature as APS (Advances Producer Services), meaning business development and supportive services. Such APS include a wide variety of services such as commercial, financial, and technical ones, but also institutional, legal and even socio-cultural and inter-cultural management ones [1, 3]. The agglomeration of APS is exceptionally high in global cities. Their combination with the various functions of global cities tends to create worldwide inter-urban networks and inter-linkages among specialized APS serving multinational corporations and their affiliates. Such linkages produce in their turn dense information exchange flows connecting global cities both in the Northern and Southern hemispheres [4].

Yet, the worldwide coverage of the global cities' literature remains somewhat limited and uncomplete. In the case of the African continent, various regions and urban centers are under-estimated. Knowledge remains insufficient in the case of semi-peripherical and peripherical cities, especially regarding their economic international functions. Furthermore, corporate internationalization through business communication inter-linkages among global cities is rarely explored [5]. This is particularly the case for developing and emerging countries such as in Africa for instance, where according to neo-institutional theories national economic indicators

---

A. Benhmade (✉) · P. Régnier · M. Spence
University of Ottawa, Ottawa, Canada
e-mail: abenhmad@uottawa.ca

and instable/risky markets prevail and do not necessarily facilitate business internationalization processes [6]. Ma [7] explain that the more foreign corporations go abroad, the more they are located in the vicinity of global cities in one mode or another, and the more their total internationalization costs tend to decrease [8]. Their smooth access to APS concentrated in such urban centers permits a substantial reduction of foreign distance, which is not only of a geographic nature.

The introductive arguments presented here above underline the central objective of this article, which is to identify a tentative conceptual framework addressing the business internationalization functions of global cities. The cross-combination of global cities' theories together with business internationalization theories lies at the very heart of this article.

15.2 Global Cities Theories

Like Katznelson (1993), we believe that the city constitutes an urban space which cannot be reduced to a single function. However, in response to the problem being faced, we confine our scope to the realm of economics, and investigate only the theories pertaining to the economic functions of global cities.

15.2.1 Theoretical Overview

It has been proven historically that cities with a global dimension have existed since antiquity in various forms, notably world-cities in the Braudelian sense. It is an economically autonomous and self-sufficient hub that exerts an influence on the factors of production within the sphere of the world-economy.

However, in the era of globalization, cities have undergone profound structural transformations which have forever reshaped their economic functions and roles. Thus, global cities are a historical construct that finds its origins in a series of events that arose especially after 1945 [9]. The emergence of global cities is defined by a plethora of changes within its ideological, political, economic and technological structures. This includes the questioning of Keynesianism, the exhaustion of the welfare state, the replacement of fordism by its successor, post-fordism, as well as the information and logistics revolution [10].

Although the concepts of world city and global city are often used to refer to cities with high levels of influence over the global economic system, they represent two different paradigms [11]. Many authors apply these two terms interchangeably without justification for a said choice [12]. The first one roots itself in the hierarchical paradigm that, according to [13], refers to the vertical organization of a space from a central city. The term "global city" finds its foundation within the network paradigm that, according to [14], describes the organization of a space by horizontal relationships among several cities. This contrast between the two concepts is not solely paradigmatic but also operational. Opting for a hierarchy-based ranking may undermine the value of a peripheral city like Jakarta, when in fact that city may well act as a coordination center to connect the global market to the Southeast Asian market.

Once a concrete definition of global cities is agreed upon, it becomes possible to explore the major theories recognized by this interdisciplinary field [1, 4, 14–21].

15.2.2 First Theories

The notion of the global city was first theorized by Sassen, in her book "The Global City: New York, London, Tokyo", in 1991.

However its inception goes back to the work of [15], for whom, the world-city only represents a conurbation.

Despite coining the term in his book "Cities in evolution", [16] is acknowledged as being the key contributor to the theory behind global cities. Whether London, Moscow, New York, Paris, Randstad, Rhine-Ruhr or Tokyo, [16] finds that all these cities exercise central power over the world-system. These consist of urban spaces where various decision-making centers cluster together, such as international organizations, public authorities, multinational enterprises, professional services, business federations, professional groups, unions, research centers, universities, museums and theaters [22]. In addition to the aforementioned institutions, major international hubs such as airports and ports put these cities at the heart of commercial and financial flows [22].

On the other hand, the work of Hall does not explain enough the link between the hierarchical world-system and global cities [23]. It is not clear whether the world-system results in the rise of global cities. In other words, in what ways does the world-system lead to the formation of global cities? To answer this, successive works have paid more attention to the world-system, and to the new international division of labor in particular.

Hymer [17] was the first to initiate the economic shift through his research on the organization of multinational firms and the hierarchical positioning of global cities. His thesis reveals that these companies tend to base their core activities within central cities, so as to be close to the decision-making centers and to establish their non-core activities in the peripheral cities [24].

Heenan [20] argues that the rise of global cities aggravates the disproportionate representation and asymmetric exchanges between the center and the periphery.

Later, Cohen [18] elucidates the links between the new international division of labor, the organizational restructuring of multinational firms and global cities. In order to oversee the execution of their decentralized operations, it is vital that these firms hire outside expertise. This governance from afar cannot be successfully established without the support of business services. These include but are not limited to, auditing-accounting, banking-insurance, management consulting, law and advertising. In addition, highly specialized services such as, certification, packaging, training, repair and transport services, are also needed.

Consequently, only global cities are revealed as appropriate commend centers with environments appropriate to foster these thriving business services [25].

15.2.3 Global City as a Command Center

The work of Friedmann [19] remains one of the cornerstones in the field. Drawing inspiration from the works by Castells [26], Harvey [27] and Wallerstein [28], with attempts to combine three major theoretical contributions, in this case, the world-system and the link between capitalism and urbanization. His aim is to highlight the functions of central cities in the post-Fordist era.

Considering the new international division of labor, the center holds a hierarchical position enabling to exercise power over the periphery. Such power tends to concentrate in central cities to the detriment of the periphery, and is unequally distributed over the space.

With the exception of São Paulo and Singapore, which are mentioned in Friedmann's ranking [19], all the global cities identified are located mainly in the Triad, meaning, North America, East Asia and the European Union.

Friedmann [19] observes that the world-system revolves around three distinct subsystems, the American subsystem, the Asian subsystem and the European subsystem. Although the American subsystem is centered on the New York-Chicago-Los Angeles axis, it encompasses a large stretch of the mainland, from Toronto in the north to Caracas in the south. Similarly, the European sub-system consists of a main London-Paris axis, and a secondary axis from the Randstad conurbation to Frankfurt and Zurich. The Asian sub-system lies on the Singapore-Tokyo axis. Lastly, the southern hemisphere is mainly centered on the Johannesburg-São Paulo axis.

15.2.4 Global Cities as Coordination Centers

Unlike her predecessors, Sassen ponders the production of resources necessary for the operation of multinational firms. By correlating between globalization, urban spaces and the tertiary sector, she concludes that global cities also act as coordination centers.

Sassen [1] points out that globalization is prompting multinational firms, known to be historically autonomous, to seek out an ever increasing number of field specific specialized services outside their headquarters. Agglomeration and geographical dispersion are at the origin, not only of the internationalization of multinational enterprises, but also of the valuation of business development services. In order for multinational firms to manage their decentralized activities, they require sophisticated services from advanced producer services, in particular, audit-accounting, banking-insurance, management consulting, law and advertising. Such professional services provide a range of strategic services, from initial bank financing to local or offshore production, marketing, advertising and final consumption.

The question is why advanced producer services tend to conglomerate in global cities like multinational enterprises? These professional services respond positively to the advantages of agglomeration due to the complex nature of their production processes, which require centrality and proximity. The non-codifiable information and the personalized services they need to deliver are all favorable conditions for their presence in global cities. From this interweaving is born a global urban network where the production of strategic services is polarized, especially in the most innovative and internationalized sectors [29].

15.2.5 Global City as a Space of Flows

In lieu of studying global cities from a hierarchical perspective, [4] relies on the network paradigm centered on the exchange of flows. Based on his research, he discovers that cities are shaped and defined by the nature of the flows that travel through them, in terms of information, knowledge, capital, goods and services and people. The higher the traffic of flows through a city, the more it becomes globalized.

Therefore, global cities are the interfaces of the space of flows which replaces the space of places. The space of flows designates several significant, repetitive and programmable series of exchanges between geographically distant actors located in global cities. These exchanges are occurring at three levels, in this case, the infrastructural level, which consolidates the technologies allowing global communication, the organizational level, which corresponds to the nodes ensuring the functioning of the networks, and the social level, which is made up of global elitist networks that transcend borders. Thus, command and coordination centers are no longer production sites, but on the contrary, global cities which act as a space of flows.

15.2.6 Global City as a Gateway to Sub-national Territory

Despite the work of Petrella [30] and Veltz [31] on global city regions, A. J. Scott is the author who is attributed for popularizing this notion [32].

Global cities are defined by their external linkages, oftentimes carried out between multinational firms and their advanced producer services. While global city regions are a product of their internal relationships, maintained by several industrial and tertiary enterprises ([33], Sassen 2001). Hence, the global city-region refers to interconnected megalopolises acting as focal points for global capitalism.

The quintessential example of the global city region is the Pearl River Delta, which connects Guangzhou, Hong Kong, Macao, Shenzhen and Zhuhai to the global economic system. While management activities are concentrated in Hong Kong and services in Guangzhou, non-core activities are dispersed throughout the rest of the delta [33].

Besides their gravitational effects in terms of the agglomeration of factors of production, global city regions fulfill key functions for the benefit of post-Fordist sectors. They act as a territorial platform, benefiting companies specializing in global finance, technological production and cultural and media industries [21]. These sectors systematically take the form of networks based on division of labor linkages [21]. They are keen to benefit from the various advantages of agglomeration, flexible specialization, diversification of activities, proximity and lower transaction costs. In addition, creativity, innovation and interconnection at sub-national, national and global scales, offer distinct benefits [2].

15.2.7 Global Cities as a Post-Industrial Production Centers

Taylor and his colleagues of the Globalization and World Cities Research network (GaWC) define global cities as post-industrial production centers organized around world-city networks [11].

A world-city network consists of three different levels: network, nodal and infra-nodal [14]. While the network level refers to the space of flows circulating in the global economy, the nodal level corresponds to cities, and the infra-nodal level to advanced producer services.

However, the question that arises is, by what processes do cities come to be connected to each other in a form conducive to the creation of a network? In response, [14] emphasizes that in spite of agreements established between urban institutions, cities are key players in the formation of their network. The main force behind economic globalization is private enterprises, which, operating in a given city, establish links with others. Advanced producer services are committed to serving their customers, regardless of the cities in which they are located [1]. As a result, the world-city network is a network of production centers that spans several cities and forms vast networks of connections on a global scale.

15.3 Business Internationalization Theories

15.3.1 First Theories

The first theories explain internationalization through the lens of market imperfections. Any firm looking to internationalize must have an advantage, which must initially be acquired in the country of origin before it can be exploited elsewhere.

Table 15.1 Economic theories of internationalization

Paradigm	Theory	Author
Structural imperfections	Initial theory	Hymer [36]
	Monopolistic advantage theory	Kindleberger [37, 38]
Transactional imperfections	English transactional theory	Buckley et Casson [39]
	American transactional theory	Williamson [40–42]
	Eclectic paradigm	Dunning [43–45]

Source [34]

Whether informational or a factor of production, a firm's specific advantage allows it to react to market imperfections. These imperfections are divided into two categories: structural and transactional ones [34]. Structural imperfections are the result of monopolistic relationships caused by privileged access to inputs, internal economies of scale, control of patented technology and control of distribution channels. Transactional imperfections refer to the idea that markets are intrinsically flawed, because information can be costly and imperfect.

All economic theories that aim to study internationalization are based on two different paradigms [35]. The first sees internationalization as the result of the internalization of financial externalities linked to structural imperfections. The second sees it as the internalization of non-financial externalities: transactional imperfections (Table 15.1).

15.3.2 New Theories

Although the early theories were the first to address internationalization, their applicability to small and medium-sized firms has some limitations [46]. They consider that firms are rational economic actors, for whom internationalization results solely from structural or transactional imperfections. In reality, internationalization is not uniquely a question of cost trade-offs for small and medium sized firms, due to limited rationality resulting from informational asymmetries.

In the early 1970s, the literature saw the emergence of a new so-called incremental theory. The stepwise theory has been the subject of many models, the best known being the so-called Uppsala model developed by Johanson and Wiedersheim (1975) [47]. It defines internationalization as a linear and evolutionary process, where the key to success is the sequential learning of the leader and their organization. As knowledge and experience of international business is acquired, the psychological distance between the company and new markets is reduced. The firm should first enter nearby markets to reduce distance, then target psychologically distant markets afterwards [47].

Model I, like the Uppsala model, also adopts a gradual reasoning. Here, internationalization refers to a process consisting of several steps, with each subsequent step being considered an innovation for the firm [48]. Typically, exporting firms evolve through a sequential process of three stages: pre-engagement, initial and advanced [49].

Although these two models are suitable for small and medium-sized companies, they are subject to limitations in the face of emerging forms of internationalization. The incremental reasoning is inappropriate for start-ups that succeed quickly or that expand operations to at least two continents in a short span of time. Johanson and Vahlne [50] propose new theories adapted to born global companies.

Hence resource and network theories suggest that firms internationalize by leveraging resources, both internal and external ones. They create networks through collaboration with their partners.

Internationalization is based on the mobilization of a set of resources known as strategic resources. These resources are rare, difficult to imitate and non-substitutable [51]. Grant [52] four attributes define strategic resources, namely transparency, transferability, reproducibility and sustainability. Penrose [53] explains that the indivisibility of productive resources results in their under-utilization, which encourages the firm to expand into other markets to take advantage of local resources. Hence the question: Can strategic resources be developed in an internalized firm-oriented way and/or an externalized network-oriented way. The selection and combination of these resources are specific to each firm according to the characteristics of their leadership, their organization and their environment [54].

Networks refer to the junction of a set of functional and relational exchanges, either in the form of inter-organizational links established between the firm and its stakeholders, or the formal and informal inter-personal links between the manager and his entourage. The more agreements a firm enters, the more access it has to the central level of its network, which allows to gain privileged access to information, to gain resources, to better control distance, and to eventually share costs [55]. Revisiting their original Uppsala model of internationalization, Johanson and Vahlne [50] emphasize the firm's position in its given network to explain its international choices. They examine internationalization multilaterally as networks developing through relationships with the host country. Networking internationalization takes the form of four sequences, namely, the evaluation of the foreign firm's positioning within its network, the identification and recognition of opportune networks in the host country, the re-examination and reconstruction of the resources to be mobilized in the host country, and, finally, the management of opportunities identified in the network [50].

15.4 Global Cities

15.4.1 Global Cities and Business Internationalization: A New Interdisciplinary Research Agenda

The notion of the global city lies at the intersection of several disciplines, including geography, urban studies, economics, development studies and sociology. It is a multi-disciplinary body of work that is rooted in six analytical perspectives, namely, historical, geographical, political, economic, social and ecological.

Historically, global cities are presented as a construct whose genesis is explained by past events [56]. Some authors rely on the so-called path dependence paradigm, according to which the past influences the present possibilities, and the present determines future possibilities. They question the historical process that led to the formation of global cities, focusing on long cycles of capitalist accumulation from the Genoese-Iberian cycle of the sixteenth century, through the Dutch and British cycles of the seventeenth and eighteenth centuries, to the American cycle of the late nineteenth century, and to the present day.

According to the geographical perspective, globalization is a phenomenon that operates from the global to the local. Authors who adhere to this approach explore the forces that lead to spatial polarization and thus to the genesis of global cities. While most of these authors analyze the global–local duality in a dichotomous way, the global exerts spatial pressure on the local, which remains passive or even inactive. Some scholars examine the multiplicity of interacting spatial scales, namely, the local, sub-national, national, regional, inter-regional and global scales. Building on this geographical perspective, other authors question the functions and roles of local forces in the development of global cities. These include states and local governments [56].

The political perspective explores the relationship between the state and global cities. Although it is accepted that globalization has led to the decline of the state and reduced its power, Couffignal [57] points out that it is more a reconfiguration of the state apparatus. The state has become a facilitator of the inclusion of cities under its control into the global economy. This raises the question of whether the state plays a role in the development of global cities. However, little research has been conducted under this political angle, since most authors adhere to the economic perspective.

Although the economic perspective has produced a plethora of theoretical and empirical postulates, all of them tend to share two core paradigms:

The hierarchical paradigm assumes that global cities exercise power over the world-system, due to their competitive positions and command functions,

The network paradigm conceives global cities as integral parts of a global urban network by means of their coordinating functions.

While the economic perspective dominates the literature, new perspectives tend to go beyond, and look at the social and ecological issues related to global cities [58].

When adopting economic lens, the focus tends to concentrate on the functions of global cities in terms of internationalization, and especially on so-called centric global cities. Such approach highlights the competitive position of the global cities of the Triad, but its analytical methodology is described as ethnocentric and subjected to some criticism. According to Robinson [59], it is a framework that is inattentive or even indifferent to the ways in which the global cities of the South are inserted into the global economic system. Although they occupy a relatively low hierarchical rank in terms of world standings, these cities are embedded in global networks. While often referred as mega-cities [60], this denomination only refers to geographical dimensions, and does not offer theoretical leverage. Beyond this descriptive classification, it remains inadequate for exploring the functions of global cities as propelled by global capitalism. Whether discussing central, semi-peripheral or peripheral global cities, all are affected to one degree or another by globalization [14]. Globalization takes place in global cities, and such cities proceed from globalization. Cross-border forces drive changes in global cities, which in turn respond to and even anticipate these changes [61].

Based on authors who have studied semi-peripheral and peripheral global cities, four economic functions are proposed namely their institutional functions, command functions, coordination functions and geographic functions.

Beyond quantitative approaches based on measurements and indices, such as the globalizing cities index, internationalization is poorly documented from a spatial perspective, especially when it comes to semi-peripherical emerging economies and marginalized developing countries. While previous research has examined the networking links among global cities, we still know little about how firms exploit such links through specific internationalization strategies and operational patterns [5]. Conventional research on internationalization tends to value the national space as the unit of analysis, even though this unit has become increasingly irrelevant precisely because of globalization [62]. As important as macro-environmental indicators can be, they are insufficient in evaluating private sector internationalization options. Foreign firms choose not only countries as investment locations, but rather, specific sub-national destinations, and micro-localization indicators are little studied. Whether it is to reduce the liability of foreignness or to obtain services and networks, firms internationalize via sub-national locations that fulfill four functions:

- An institutional ecosystem favorable to the reduction of institutional distance,
- A destination offering a high density of supportive services and networks,
- A location connected to other locations nested in global urban networks,
- A specific gateway that facilitates business access to distant and/or neighboring markets.

Therefore, it seems likely that a company can benefit from more location advantages when it transits through a well positioned global city from hierarchical and networking viewpoints.

15.4.2 Theoretical Framework Search

Building on global cities and internationalization theories, our search pleads for an interdisciplinary conceptual framework. It assumes that global cities can not only act as specific ecosystems conducive to foreign business internationalization combined with a dense localization of business development services and networks connecting national, regional, inter-regional, continental, inter-continental and global markets (Fig. 15.1).

On the supply side, global cities concentrate business services and networks to facilitate internationalization to all types of regional destinations and beyond. On the demand side, firms seek new resources (services and networks) to reduce distance and liability of foreignness. On the one hand, internal resources refer to human resources and their knowledge and skills in international operations and transactions, to financial resources, as well as to the technical resources necessary to create a specific competitive advantage. On the other hand, external resources refer to a set of

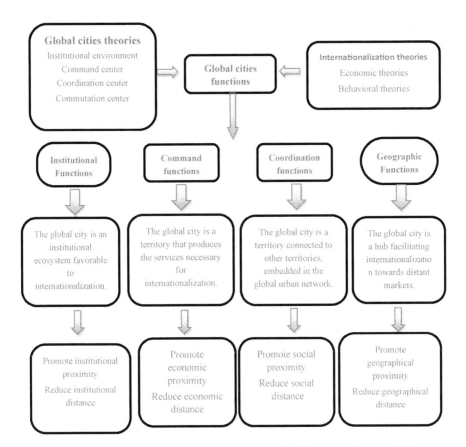

Fig. 15.1 Proposed theoretical framework. *Source* Benhmade, Régnier and Spence, 2023

networks providing access to public, para-public and private business development services, which provide a wide range of financial and non-financial support. Such networks pass through concentration nodes often located in global cities. However, the optimal use of these resources depends largely on the company and its team.

15.4.3 Global Cities Functions

Institutional Functions

The global city seems to provide an institutional ecosystem particularly adapted to foster internationalization. Any foreign company wishing to expand its activities abroad is confronted with institutional distance and complexity all the way. The greater the institutional incompatibility between home and host ecosystems, the more institutional differences constrain the decision to internationalize. Under such conditions, foreign firms favor less costly modes of business entry. Eden and Miller [63] find that in institutionally distant countries, foreign firms face greater liability of foreignness [64]. These are not economic costs related to local production or distribution constraints, but costs related to uncertainty, discrimination and complexity [65]. While uncertainty refers to the lack of knowledge and experience in foreign markets of destination, discrimination results from unfavorable treatment of foreign firms by local stakeholders. In addition, complexity is caused by the difficulty of managing business relationships, whether intra-organizational or inter-organizational [5]. It is possible to overcome such institutional distance and complexity, especially when it comes to business internationalization in emerging and developing countries. Although such market destinations are generally undeveloped from an institutional perspective, peripherical global cities are capable of performing institutional functions reducing distance and complexity, and provide business proximity and access facilitation services [66].

Command Functions

The global city can also act as a producer or provider of services supporting business internationalization. Any foreign company going international, even a multinational firm, needs a range of business development services to access local governance and markets. Such services are generally some provided by public institutions, namely, motivational, informational and operational ones [67], as well as private services offered by producers of strategic services, such as auditing-accounting, banking-insurance, management, law and advertising [1]. They range under the so-called denomination of Advanced Producers Services (APS). The higher the recurring and non-recurring costs of internationalization, the greater the incentive for a foreign company to obtain support from external services locally. This is especially true in emerging and developing countries, because of the geographic, institutional, intercultural and even psychological distance leading to high or even prohibitive transaction costs. Therein lies the advantage of the agglomeration of service producers concentrated in global cities. Such APS providers reduce the gap between the country of origin and the country of market destination. The local presence of such services prevents from the need to import them from the foreign country of origin and allows frequent collaboration with the same types of APS providers around the world [5].

Coordination Functions

The global city can also function as an urban cener connected to other urban centers, nested with each other in a global urban network continuum [1, 68]. In the era of the New Economy, which is at once global, informational and reticular, companies are increasingly encouraged to organize themselves through global networking [69]. The more fragmented the division of labor, along complex value chains the greater the need for networking coordination locally and globally. Different forms of distance as underlinesd above lead to various types of business informational asymmetry, and therefore hinders the development of business and management relationships [70]. According to [2], by enabling coordination among distant foreign and local actors and de-territorialized business activities, global cities facilitate the circulation and transfer of information, goods, services, capital and people in fast and non too costly manners. More precisely, producers of APS services, present across the globe, manage to create inter-urban linkage servicing flows. Although the informational revolution has favored the circulation of intangible flows, strategic decisions can only be made when firms meet [71]. Exchanges that are sensitive to distance and transaction costs can only occur in spaces that are conducive to business flows' agglomeration [72]. Such agglomerations are enabled through the development of relational and functional networks. While such relational networks are essential to feed functional networks, the latter facilitate access to APS services facilitating foreign operatons in distant markets.

Geographic Functions

Global cities function as central hubs to access neighbor and regional markets. They constitute nodes through which local, sub-national and national economies are interconnected with both local markets and the global economic system [69, 72]. Although semi-peripheral and peripheral global cities have little influence over global capitalism, they are nonetheless important to its circulation from centers to peripheries and vice versa. The gravitational weights of global cities impacting the expansion of foreign firms depend on their motives for exports and investment overseas. Building on the taxonomy presented by Behrman [73] and Dunning [74], Goerzen et al., [5] suggest that firms seeking new markets are more likely to use global cities as gateways, than firms seeking strategic assets, resources or efficiency. Unlike management activities, which are often polarized in OECD economies, execution activities are often internationalized towards emerging and developing countries. This internationalization is not aimed at creating new upstream skills, as is the case with knowledge-intensive activities such as research and development, but rather to exploit downstream skills such as production and distribution. Through the use of their node-switching functions, often both local and global in scope, global cities act as physical hubs opening access to geographically distant markets and expanding customer portfolios. Through their interconnectivity between distant and nearby markets, global cities can be particularly attractive to foreign companies wishing to set up so-called global market servicing production and distribution networks. This is not only a question of logistical means, but also of informational strategies which operate as promotional channels on a large geographical scope and scale.

15.5 Futures Avenues

The above proposed conceptual framework may help to answer the core research question whether global cities facilitate business internationalization. This main question can be divided into four subsidiary questions, namely:

(1) Do global cities offer a proper business environment for the development of international business?
(2) Do they propose APS services facilitating business internationalisation?
(3) Do they enable foreign firms to connect with local stakeholders in the national/regional market of destination? And
(4) Do they play a central business networking role to connect foreign firms to neighbor and more distant markets regionally and beyond?

15.5.1 Epistemological Paradigm

Eisenhardt [75] distinguishes between the testing of existing knowledge and the generation of new knowledge. The former permits the evaluation of validity and reliability of existing theories, the latter the tentative production of new theories. The issue is therefore to explore the functions of global cities in terms of business internationalisation from the theoretical perspective of global cities combined with international business management theory. However, as few research works have focused so far on the functions of global cities located in emerging and developing countries, it is suggested here to generate some knowledge by studying the case of Casablanca, and to focus more specifically its global city facilitation role it may provide to Canadian firms expanding into early twenty-first century African frontier markets. Collecting and testing such empirical knowledge may lead to further enrichment of exsiting knowledge ultimately sheding light on identified research holes and gaps. Originally, global cities theories were mainly based on the specificities of cities in developed countries and their roles vis a vis multinational firms. The application of such theories to global cities in emerging and developing countries requires an evaluation that goes beyond existing collected empirical knowledge. A mixed research approach combining deductive and inductive reasoning may be appropriate to meet such epistemological objectives.

15.5.2 Research Methodology

From an empirical perspective, a qualitative research can be used. The nature of the data to be collected in a case study such as the global city of Casablanca provides little support for a quantitative enquiry. A qualitative enquiry seems appropriate to explore and document the functions filled by a semi-peripherical global city such as Casablanca located in an emerging economy such as Morocco. Each function and corresponding support APS service needs to be studied individually and according to the needs of each individual Canadian firm in a proposed representative research sample of Canadian firms expanding into African markets through a global city such as Casablanca. It must be specified here that only three leading global cities are ranked on the African continent, namely Cairo and Johannesburg in addition to Casablanca.

Based on a proposed case study of Canadian firms using possibly various global city functions concentrated in Casablanca, these functions can be studied from two perspectives, namely the supply of APS support services established in Casablanca, and the demand as expressed by Canadian firms already present or expanding into African markets. With the help of multiple data collection tools and data triangulation, this case study approach allows to reach two different samples:

> The APS service providers that support foreign firms that transit through Casablanca.
> The Canadian firms using such services to expand their international business reach to Africa.

Several intra-case case comparisons can be conducted as well as inter-cases. While the intra-case study presents each case in the sample, inter-case studies help to compare Canadian firms within the two selected samples. To these two inter-case studies will be added an inter-sample study with the objectives of triangulating answers from the two samples, and comparing and contrasting the results.

15.5.3 Research Sampling

Respondents have to meet a range of pre-defined criteria [75]. Given informational business management research needs, two distinct samples can be constructed.

The first sample can include providers of strategic services based in Casablanca, and whose mission is to support foreign firms that have internationalised in Africa. This sample has to meet the four following criteria: geographical location, sectoral profile, type of supplied APS service, and level of business internationalization involvement. Such an intersectoral sample needs to sample providers of financial and non-financial services, audit, banking and insurance, management consulting, legal services and advertising.

The second sample can include Canadian firms expanding into African markets, selected according to five sampling criteria, namely localisation, market targets, business sector, business international involvement and informant. Canadian firms refer to companies being primarily registered in Canada.

For appropriate and pertinent sampling, the number of Canadian firms to be included can be decided not ahead of time, but during the research process based on reply and similarity replication so that scientific research reliability is ensured (Yin 2009). Saturation will be reached once each Canadian firm case study and corresponding APS support services will not bring any new empirical knowledge. The recruitment of between four and ten respondents per sample should help to reach saturation [75].

15.5.4 Data Collection and Processing

Two data collection techniques cab be used independently or combined depending on the cases: non-intrusive documentary research and semi-directed interviews.

Before contacting respondents, it is important to collect and analyse various data such as factual information on Casablanca business environment, the attractivity of its ecosystem to foreign firms, and the policies deployed to promote its global city ranking further and further compared to its African competitors. Some core research questions need to address the following themes: geographical location, institutional advantages, attractivity related to business networks.

Four methods of data analysis can be used: cognitive, lexical, linguistic and thematic methods. Thematical analysis may be privileged as it facilitates the treatment of collected data according to pre-established themes.

Data analysis will be divided in four levels of analysis: intra-case, inter-case, inter-sample and theoretical levels. The intra-case analysis is to analyse each case individually. The inter-case analysis enables comparison within the two samples. The inter-sample analysis combines the two samples and compare the data of both. Finally, the theorising analysis may lead to the theoretical framework and to data triangulation [76].

Hence, through deduction it could valid some hypothetical propositions, and inductively identify new findings, and suggest by abduction other research paths forward.

15.6 Conclusions

As central urban center incorporating major economic functions, global cities act as business hubs contributing to a worldwide economic system. In the economic sphere, global cities concentrate business development services and facilitation networks, and serve as global and interconnected crossroad nodes.

Early research since the 1990s and early 2000s focused on key global cities such as the financial triad of London, New York, and Tokyo. However, although the empirical coverage of original literature on the subject has remained somewhat limited so far beyond quantitative research rankings. Various emerging global cities are underestimated or even neglected, and African ones in particular. Even if they occupy relatively low hierarchical positions in the rankings of global cities, they are anchored in the global urban network which reaches practically every market location on the planet. More recently, research has been completed on some other global cities, but mainly in OECD economies and a few elsewhere on the three emerging continents. While classical literature has so far been interested in so-called central global cities, new research admits that emerging and developing cities are also part of the global economic system. This observation becomes true when exploring the economic functions of emerging global cities, namely their institutional functions, command functions, coordination functions and of course their geographic functions. In the case of Africa, the rapid emergence of Lagos as a fourth global city is a good illustration in addition to Cairo, Casablanca and Johannesburg.

This chapter has aimed to cross fertilize global cities theories from an economic perspective together with business management internationalization studies. The research contribution aims to explore and document the entrepreneurship support functions of global cities addressing and facilitating the worldwide territoriality expansion of corporate internationalization. Such internationalization proceeds from multifactorial and multichoice parameters. On the empirical front, practical oriented research pathways are suggested to identify and study foreign firms, in particular their whys and hows to choose global cities' multiple functions and use the wide diversity of highly skilled APS support services more specifically. In the case of Canadian corporations, over 75% of their total exports serve the US market, whereas less than 1% go to Africa. However, the African continent is described as the new and last global business frontier of the twenty-first century. Global cities such as Casablanca are foreseen as playing a key business entry gateway role in this continental context.

References

1. Sassen S (1991) The global city: New York, London, Tokyo. Princeton University Press, Princeton
2. Camagni R (2001) The economic role and spatial contradictions of global city regions: the functional, cognitive and evolutionary context. In: Scott AJ (ed) Global city-regions: Trends, theory, policy. Oxford University Press, Oxford, pp 96–118

3. Taylor P, Derudder B, Faulconbridge J, Hoyler M, Ni P (2014) Advanced producer service firms as strategic networks, global cities as strategic places. Econ Geogr 90(3):267–291
4. Castells M (1999) The information age: economy, society and culture. B. Blackwell, Oxford
5. Goerzen A, Nielsen BB, Asmussen CG (2013) Global cities and multinational enterprise location strategy. J Int Bus Stud 44(5):427–450
6. Kittilaksanawong W (2017) Influence of region, country and subnational-region institutions on internationalization of multinational corporations. In: Munoz M (ed) Advances in Geoeconomics. Routledge, New York, pp 46–54
7. Ma X, Tong TW, Fitza M (2013) How much does subnational region matter to foreign subsidiary performance? evidence from fortune global 500 corporations' investment in China. J Int Bus Stud 44:66–68
8. Nachum L (2003) Liability of foreignness in global competition? financial service affiliates in the city of London. Strateg Manag J 24:1187–1208
9. Boschken HL (2011) Global cities. In: Southerton D (ed) Encyclopedia of Consumer Culture. SAGE Publications, Thousand Oaks, pp 664–665
10. Warf B (2006) Global cities. In: Warf B (ed) Encyclopedia of Human Geography. SAGE Publications, Thousand Oaks, pp 195–196
11. Taylor P, Ni P, Derudder B, Hoyler M, Huang J, Witlox F (2011) Global urban analysis: a survey of cities in globalization. Routledge, London
12. Derudder B (2006) On conceptual confusion in empirical analyses of a transnational urban network. Urban Studies 43(11):2027–2046
13. Arrault J-B (2006) L'émergence de la notion de ville mondiale dans la géographie française au début du xxe siècle. Contexte, enjeux et limites. L'Information géographique 70(4), 6–24.
14. Taylor P (2004) La régionalité dans le réseau des villes mondiales. Revue internationale des sciences sociales 131(3):401–415
15. Geddes PS (1915) Cities in evolution : an introduction to the town planning movement and to the study of civics.
16. Hall P (1966) The world cities. Weidenfeld and Nicolson, London
17. Hymer S (1972) The multinational corporation and the law of uneven development. In: Bhagwati JN Economics and world order-From the 1970's to the 1990's. The Macmillan Company, London
18. Cohen RB (1981) The new international division of labour, multinational corporations and urban hierarchy. In: Dear M, Scott AJ (eds) Urbanization and urban planning in capitalist society. Metheen, London, pp 287–315
19. Friedmann J (1986) The world city hypothesis. Dev Chang 17(1):69–83
20. Hennan DA (1977) Global cities of tomorrow. Harv Bus Rev 55:79–92
21. Scott AJ (2001) Global city-regions: trends, theory, policy. Oxford University Press, Oxford
22. Clark D (2003) Urban world/global city. Routledge, London
23. Brenner N (1998) Global cities, glocal states: global city formation and state territorial restructuring in contemporary Europe. Rev Int Polit Econ 5(1):1–37
24. Beaverstock JV, Smith RG, Taylor PJ (1999) A roster of world cities. Cities 16(6):445–458
25. Acuto M, Steele W (2013) Global city challenges: debating a concept, improving the practice. Palgrave Macmillan, Basingstoke
26. Castells M (1972) La question urbaine. F. Maspero, Paris
27. Harvey D (1973) Social justice and the city. Edward Arnold, London
28. Wallerstein I (1979) The capitalist world-economy. Cambridge University Press, Cambridge
29. Sassen S (2004) Introduire le concept de ville globale. Raisons Politiques 15(3):9–23
30. Petrella (1995) A global agora vs. Gated city-regions. New Perspect Q 21–22
31. Veltz P (2014) Mondialisation, villes et territoires: l'économie d'archipel. Presses Universitaires France, Paris
32. Derudder B, Witlox F (2008) What is a « World Class » city? comparing conceptual specifications of cities in the context of a global urban network. In: Jenks M, Kozak D, Takkanon P (eds) World Cities and Urban Form. Routledge, New York, pp 11–24

33. Hall P (2001) Global city-regions in the twenty-first century. In: Scott AJ (ed) Global city-regions: Trends, theory, policy. Oxford University Press, Oxford, pp 59–77
34. Sierra C (2003) Firmes multinationales et dynamique de création technologique : Des coûts de transaction aux coûts de coordination. Université Lyon 2, Lyon.
35. Hennart J-F (1991) The transaction cost theory of the multinational enterprise. In: Pitelis C, Sugden R (eds) The Nature of the Transnational Firm. Routledge, London, pp 81–116
36. Hymer S (1960) The international operations of national firms: a study of direct foreign investment. MIT Press, Cambridge
37. Kindleberger CP (1969) American business abroad: six lectures on direct investment. Yale University Press, New Haven
38. Caves R (1971) International corporations: the industrial economics of foreign investment. Economica 38(149):1–27
39. Buckley PJ, Casson M (1991) The future of the multinational enterprise. Macmillan, Houndmills
40. Hennart J-F (1982) A theory of multinational enterprise. University of Michigan Press, Ann Arbor
41. Williamson OE (1975) Markets and hierarchies. Challenge 20(1):70–72
42. Teece DJ (1983) Technological and organisational factors in the theory of the multinational enterprise. In: Casson M (ed) The Growth of International Business. George Allen & Unwin, London, pp 51–62
43. Dunning J (1979) Explaining changing patterns of international production: in defence of the eclectic theory. Oxf Bull Econ Stat 41(4): 269–295.
44. Dunning J (1980) Toward an eclectic theory of international production. Int Exec 22(3):1–3
45. Dunning J (1988) The eclectic paradigm of international production: a restatement and some possible extensions. J Int Bus Stud 19(1):1–31
46. Allali B (2005) Vision des dirigeants et internationalisation des PME: une étude de cas multiples de PME marocaines et canadiennes du secteur agroalimentaire. Publibook, Saint-Denis
47. Johanson J, Vahlne J-E (1977) The internationalization process of the firm-a model of knowledge development and increasing foreign market commitments. J Int Bus Stud 8(1):23–32
48. Gankema HGJ, Snuif HR, Zwart PS (2000) The internationalization process of small and medium-sized enterprises: an evaluation of stage theory. J Small Bus Manage 38(4):15–27
49. Leonidou LC, Katsikeas CS (1996) The export development process: an integrative review of empirical models. J Int Bus Stud 27(3):517–551
50. Johanson J, Vahlne J-E (2009) The Uppsala internationalization process model revisited: from liability of foreignness to liability of outsidership. J Int Bus Stud 40(9):1411–1431
51. Barney J (1991) Firm resources and sustained competitive advantage. J Manage 17(1):99–120
52. Grant R (1991) The resource-based theory of competitive advantage. California Manage Rev 33(3):114–135
53. Penrose E (1959) The theory of the growth of the firm. John Wiley & Sons, New York
54. Laghzaoui S (2009) Internationalisation des PME: apports d'une analyse en termes de ressource et compétences. Mana Aveni 2:52–69
55. Forsman M, Hinttu S, Kock S (2002) Internationalization from a SME Perspective. In: IMP Conference. Présenté à Dijon. Dijon
56. Davis D, Santamaría G del C (2010) Global city. In: Hutchison R (ed) Encyclopedia of urban studies. SAGE Publications, Thousand Oaks, pp 314–316
57. Couffignal G (1997) Le rôle de l'État en Amérique latine: pistes de recherches. Cahiers des Amériques Latines 26:83–191
58. Jenks M, Kozak D, Takkanon P (2008) World cities and urban form. Routledge, New York
59. Robinson J (2002) Global and world cities: a view from off the map. Int J Urban Reg Res 26(3):531–554
60. Dogan M, Kasarda J (1988) A world of giant cities: the metropolis era. SAGE Publications, Thousand Oaks
61. Short JR, Kim Y (1999) Globalisation and the city. Longman

62. Brown E, Derudder B, Parnreiter C, Pelupessy W, Taylor P, Witlox F (2010) World city networks and global commodity chains: towards a world-systems' integration. Global Netw 10(1):12–34
63. Eden L, Miller SR (2004) Distance matters: liability of foreignness, institutional distance and ownership strategy. Adv. Int. Manag. 16:187–221
64. Zaheer S (1995) Overcoming the liability of foreignness. Acad Manage J 38(2):341–363
65. Zaheer S (2002) The liability of foreignness, redux: a commentary. J Int Manage 8:351–358
66. Martinus K, Sigler TJ (2018) Global city clusters: theorizing spatial and non spatial proximity in inter-urban firm networks. Reg Stud 52(8):1041–1052
67. Czinkota M (2002) Export promotion: a framework for finding opportunity in change. Thunderbird Int Bus Rev 44(3):315–324
68. Taylor P (2001) Specification of the world city network. Geogr Anal 33(2):181–194
69. Castells M (1991) The information age: economy, society and culture. B. Blackwell, Oxford
70. Bergen M, Dutta S, Walker O (1992) Agency relationships in marketing: a review of the implications and applications of agency and related theorie. J Mark 56(3):1–24
71. Törnqvist G (1970) Contact systems and regional development. Gleerup, Lund
72. Scott AJ (2001) Globalization and the rise of city-regions. Eur Plan Stud 9(7):813–826
73. Behrman JN (1972) The role of international companies in Latin American integration. Lexington Books, Lexington
74. Dunning J (1993) Multinational enterprises and the global economy. Addison-Wesley, Boston
75. Eisenhardt K (1989) Building theories from case study research. Acad Manag Rev 14:532–550
76. Robert D, Gaudet S (2018) L'aventure de la recherche qualitative: Du questionnement à la rédaction scientifique. Presses de l'Université d'Ottawa, Ottawa
77. Albaum G, Duerr E (2008) International marketing and export management. Pearson.
78. Baldegger RJ, Wyss P (2007) Born-again global firms in Switzerland: a case study on internationalization behaviour of established firms, Fribourg, Bern, Newyork.
79. Catanzaro A (2014) Influence des Services d'Accompagnement à l'Export sur les ressources et la performance internationale des Exportatrices Précoces [Thèse]. Université Montpellier 1.
80. Julien PA (1994) Les PME: bilan et perspectives. Economica, Paris
81. McDougall P, Oviatt B (2000) International entrepreneurship: the intersection of two research paths. Acad Manag J 43(5):902–906
82. Ruzzier M, Hisrich R, Antoncic B (2006) SME internationalization research: past, present, and future. J Small Bus Enterp Dev 13(4):476–497
83. Seringhaus FHR, Rosson PJ (1991) Export development and promotion: the role of public organizations. Kluwer.
84. Régnier P, Wild P (2018) SME internationalisation and the role of global cities: a tentative conceptualisation. Int J Export Mark 2(3):158–179
85. Yin RK (2009) Case study research: design and methods. Sage Publications, Thousand Oaks.
86. Zucchella A, Scabini P (2007) International entrepreneurship: theoretical foundations and practices. Palgrave Macmillan.

Printed in the USA
CPSIA information can be obtained
at www.ICGtesting.com
LVHW020812170924
791295LV00003B/234